VALOR™

*The Ultimate Handbook on Military
Professional Transition to
Your Next Career!*

By: Kevin Preston, MBA (COL-Retired) &
Dr. Jeffrey Magee, CMC/CBE/CSP/PDM

ISBN: <u>979-8-9892760-2-8</u>

US $29.99

Copyright 2024

Preston, Kevin, MBA (COL-Retired)

Magee, Jeffrey L., PhD/CMC/CBE/PDM/CSP

Valor ™: The Ultimate Handbook on Military Professional Transition to Your Next Career!

Green, Sheryl, editor

Braster, Jennifer, copy editor

Peak Image Designs, cover design editor

Performance 360 Media Group/ Jeffrey Magee, LLC, publisher

Performance 360 Media Group Las Vegas, Nevada

www.jeffreymagee.com

A New Era in Publishing™

For information regarding special discounts for bulk purchases for business training emersion, large groups, personalized edition for your group, families and gifts, please contact the following:

Las Vegas, Nevada

<u>www.JeffreyMagee.com</u>

Foreword

The transition from the military to civilian life is hard. I know as I served over 40 years both in the National Guard and on active duty. Recently, I, too, transitioned from active to retired. This self-guided brilliant book leads you to understanding your value, finding purpose, and reconciling your wants and needs. All of those objectives matter. These two superb leaders provide us with the VALOR system as a roadmap to success personalized by you. Most importantly, you can trust that this system works.

As I culminated my career in the US Army, I served as the Training and Doctrine Commander (TRADOC), where I was ultimately responsible for recruiting, training, educating, and transitioning over 800,000 Soldiers a year. I commanded in combat six times and spent the bulk of my career with combat units with our nation's warriors. My culminating wartime effort was being the three-star in charge of the wars in Iraq and Syria.

General Paul Funk

For More on Paul CONNECT-
Gen. Paul E. Funk II (Ret) | LinkedIn Link:
https://www.linkedin.com/in/paul-e-funk-ii/

CONTENTS

Introduction

About The Authors

Preston –

As a career Army officer, I spent two-plus decades watching soldiers end their terms of service or retire, only to see them struggle with life following service, particularly with employment. As a young officer, I thought there had to be a better way.

Soldiers that I admired, who were brilliant, confident, extremely well educated, had an unwavering moral compass, and commanded the respect of their superiors and peers somehow struggled with employment. Upon departure, they often lost their identity, their network of friends, and their routine simply ceased to exist. Losing that military "rudder" was devastating.

I knew there had to be a better route, a path by which a transitioning service member could gain the kind of employment they desire, one that fills some of the void of military service.

Early in my quest, I established three personal goals:

1. I want to depart military service on my terms. I had a glorious career, and I did not want to be "forced" out due to a mandatory removal date. That would simply tarnish a great career.
2. I did not want to work as a federal, state, or military contractor.
3. I wanted my next career to be something different.

This book is a chronology of the six years I spent learning how to leave service in the Army and gain employment as a senior leader with The Walt Disney Company. I achieved all three of the above goals. Looking back on this, I realized that what I did was not exceptional. I applied the skills and tools taught to me by the Army, became a humble student of learning how to brand and market myself, and took charge of my future following 28 years of service.

What came from this 6-year learning journey was a series of steps, make no mistake it was not a well-organized plan while I was in the process. Rather it was trying many things to see what would work. What worked became this book, "Valor."

For More on Kevin, CONNECT -
Kevin Preston, MBA, M Ed
Linked In: https://www.linkedin.com/in/kevin-preston-mba-m-ed-9a92555/

Magee –

Through decades of employment, downsizing, rightsizing, business growth, and executive suite positions, charting one's career trajectory and next career is something I have experienced first-hand.

Through my talent development firm, www.JeffreyMagee.com, I have had the opportunity to design the talent development programs used by the military and private sector for decades. Chronicled through 30 books, translated into 21 languages, to include graduate management textbooks and four global bestsellers, understanding how to frame your talent and align it with winning organizations is what we/I do.

Combined with my media firm, for decades, I have been an employee and employer. I have served as the Publisher/Editor-in-Chief to the global publication, PERFORMANCE/360 magazine aka www.ProfessionalPerformanceMagazine.com, which has aligned me with the top branded personalities, employers, and military senior NCOs and Generals. I've spent decades as a business writer, best-selling author of graduate management text and trade books, business owner, and signer of lots of personnel paychecks, along the way becoming a talent development advisor. In my own career, I have had the opportunity to transition from one employer to another, both by design and by forced initiatives. I have worked with military professionals in their transition campaigns for decades. I have seen and heard everything there is to witness around the topic here of talent management and how one does and does not successfully transition from the military to the private sector for sustained next-chapter success!

I have served and I have observed from massively successful people and organizations what works and what does not work.

As a professional business genre journalist and broadcaster for many years in the Midwest, I transitioned into the world of the Fortune 100 in a selling and business capacity. Through the years, I've built myself into working with senior military leaders and business executives, penned 31 books, translated into 21 languages, along the way. With my work with the Fortune 100 and cutting-edge entrepreneurial businesses and for nearly three decades with JeffreyMagee.com, a leadership talent development firm, I have also had the unique opportunity to collaborate with the top two employment search/placement firms CEOs in America – Max Messmer of Robert Half Inc. and Robert Funk of Express Employment Professionals.

This book is a chronology of 30 years of front-row seats to military leaders and private sector business professionals' trajectory to continued success.

For More On Jeffrey CONNECT –

Dr. Jeffrey Magee Business Development/Leadership/ Author

Linked In: https://www.linkedin.com/in/drjeffspeaks/

www.JeffreyMagee.com

www.ProfessionalPerformanceMagazine.com

The VALOR System

Throughout this book, we will map out The **VALOR System,** a plan-of-action for you to use as you transition out of service. *VALOR: The Ultimate Handbook on Military Professional Transition to Your Next Career!* will guide you through strategies and tactics to raise your **VISIBILITY**, illustrate ways for you to gain **AWARENESS** in the marketplace, and show you how to **LEVERAGE** yourself for your next **OPPORTUNITY** while being strategic in the **RELATIONSHIPS** you have now and in your later career.

How To Use This Book

As you read this book, take note of where you are in your career ladder and where you are in your strategy or execution for career transition. If you are a veteran, reflect upon where you are and where you want to be in the coming years.

At the end of each Chapter, there is a **CALL TO ACTION (CTA) PAGE**, along with a space to take notes. Build your transition game plan here as you read and as you go.

Add Key Performance Indicator CTAs into your calendar to serve as your personal coaching reminder to action.

And engage your mentors, sponsors, advisors, and coaches as we will share throughout this book to serve as your own force multipliers!

Chapter 1

Accept

Your military career will eventually come to a close.

No matter how long or how well you have served, this is an undeniable fact. This may bring up a storm of emotions, but you needn't encounter anxiety of the unknown. Just as you do daily within your military career, you take calculated actions to survive and thrive and that is the same focus in your career transition and in securing your next career opportunity. With the help of the VALOR System, you can experience the excitement of the "what's next" opportunity that you have strategically worked towards. With thousands of veterans entering the private sector every month, atop an already stunningly high number of already-veterans participating in the workplace, you must stand out and ensure you bring massive value and receive an open-door reception. Thankfully, this is much easier than you may think.

To make the move from a career in the armed forces to a civilian career, you must first accept that your military career is coming to a close. You must **accept** the closure of that career and that phase in your life. Only when you embrace that your military career is truly closing, will you take the steps to prepare for the next career.

Military service is one of the most immersive careers a person can have. Consider this: in what other career can you work, shop, worship, travel, and live in one community? That is service in the armed forces. This is not only a job with a paycheck, but often a calling, and a person's identity is tied directly to military service.

Asking a service member to simply drop that lifestyle, identity, and purpose is a tall order. It is much, much different than a civilian moving from one company to another.

Value Your Experience

Military values, regardless of branch, are a strong driver of behavior. While many of the words are different among the branches, a consistent theme revolves around service, duty, honor, and loyalty. This ethos produces a service member who will focus on their military job at the exclusion of everything around them.

While serving, you demonstrate and exercise loyalty to your comrades, your chain of command, the mission and purpose of your unit, the nation, and the constitution. Do not let this value of loyalty become an anchor that holds you back from moving to the future. We have lived loyally throughout our years of service.

When you leave the military, you can continue to exercise that loyalty by becoming an example for others of what a great transition looks like. You will not only inspire and teach, but make an impact on countless other veterans. Your loyalty does not evaporate upon accepting your DD214, instead, you will exercise loyalty to yourself, family, and other service members as they transition.

You should be excited and proud of your military career and accumulated accomplishments. You are employed right now by the #1 business on the planet engaged in on-going training, informal education development, formal accredited education, certifications galore, job development, execution recognition opportunities, and continual consistent career promotion advancement opportunities. Chronicle all of the education, certifications, licenses, volunteer affiliations, and experiences one has when they near retirement days. The private marketplace is hungry for you!

Accept and Embrace the Change

How do you learn to accept such a significant shift in your life?

First, view yourself as your own entity from day one: YOU, Inc. Think like your own brand; you are the entrepreneur of your own life and where it goes. You may want to ensure you have a robust LinkedIn profile and if available, buy your name as a URL domain name. We will discuss this in depth in Chapter 7.

Second, understand that you are dispensable. So many of us believe "If I am not here, my job will not be done!"

We must understand that none of us are indispensable. The United States Armed Forces has been around for 200+ years and the departure of any single individual will not stop the system. It's time to let go and look toward your future post-military life.

Third, use time as an ally, not an enemy. Taking plenty of time to plan a transition (no less than 2 years and up to 5 years) will help a service member slowly walk in a planful and deliberate fashion into their next career.

Begin Your Transition

We can look at your military service in terms of seasons. Please note, whatever season you find yourself in is the right time to start planning for your exit.

Spring – Your entry point and on-boarding phase to the military. The fundamentals are established, and your first duty assignment under your career job titling is assigned. If you think through this information and look at the greatest achievers within these ranks, you can emulate their behaviors and follow their accelerated success. Most military professionals never think about the who, what, when, where, why, and how of building a great life and military career this soon in their career, but this is where the seeds to your military transition actually start. You can plan backwards from your separation date with milestones and achievements, including those associated with service. My ARMY backwards consisted of:

1. Upskilling and education
2. Many rounds of resume reviews
3. Mentoring from business professionals
4. Reading and digesting association and business periodicals
5. Learning how to speak "civilian" and the language of business

These are just a few of my pre-transition milestones. Approaching my retirement this way gave me something tangible to bite into and helped "soften" the blow and accept that this career was closing.

Summer – This is the season of aptitude development and gain, the season of application, doing, achieving, learning, perfecting, and experiencing. Be hungry to grow and explore everything that makes sense. The earlier into your military experience that you can be strategic about your assignment, schooling, education, certifications, and networking, the better your transition will be!

Fall – This is where you start to see the fruit of the previous season of work, successes, and tribulations. During this time, you will:

- Double down on what works while adjusting what does not. Move forward while assuming extreme ownership.

- Forge your own branded reputation within organizations and among others.

- Create powerful and lasting relationships.

- Recognize those who are in it for themselves and have created a brand that others view with disdain. Don't become one of them; this will potentially be speed bumps and cliffs to your transition success.

Winter – This is where your career is winding down or the career pathway you've been on is becoming blocked. Here is where serious calculated work accelerates. Sometimes a horizontal movement within your military assignments may be needed to put you into a better vertical trajectory for a better transition opportunity. Have you explored all of the resources, programs, and transition professionals that the military has available? There are endless on-line resources, portals, and agencies you can consider to expose you to civilian opportunities.

If you operate this way from day one with some sort of curiosity as to what industries you may want to work in post-military, you will be more strategic in:

1. MOS, AFSC selection(s)

2. Where geographically you want to go for your next assignments (think globally here and not just domestically) and consider how this can be leveraged in the future into your network/relationships

3. What educational opportunities, experiences, certifications, degrees, schooling, and programs excite you now and will have leverage in your next life, double-down on them

4. Recognize the professionals you may meet and can purposefully connect with to enhance your journey and their journey now as this will be valuable later.

At a minimum, you should start putting your actual transition plan into play within five years of your separation date. The anxiety rises as most active duty professionals at every rank tend to play a game of blitzkrieg in their final 18 months.

If you are not much of a reader, become one now. If you like to read, become more strategic. Here are a few books that need to be on your professional development bookshelf:

- ***How to Win Friends and Influence People™***, first appeared in 1936 as a personal and professional self-help book written by Dale Carnegie. Over 30 million copies have been sold worldwide making it one of the best-selling books of all time.

- ***The 7 Habits of Highly Effective People™***, first published in 1989, is a business and self-help book written by Stephen R. Covey.

- ***YOUR TRAJECTORY CODE: How To Change Your Decisions, Actions, and Directions, to Become Part of the Top 1% High Achievers™***, first published in 2015, is a business professional's ultimate handbook to success by Jeffrey Magee.

In the next chapter, we will help you envision your future.

Call-to-Action:

Now reflect on what you have just read, ideas that have been activated in your head, and weigh them against each element of the **VALOR™ Model** and what you need to be doing right now at this moment in your career/professional life and with what makes sense for you ...

VALOR: The Ultimate Handbook on Military Transition

What are you doing and need to be doing with respect to the strategies and tactics to raise your **VISIBILITY**,

_____

_____

_____

Illustrate ways for you to gain **AWARENESS** into the marketplace,
How can you **LEVERAGE** yourself for your next **OPPORTUNITY**,

_____

_____

_____

While being strategic in the **RELATIONSHIPS** you have now, who are they,

_____

_____

_____

_____

And who can you identify that you don't know and need to get connected to for potential advocacy in your later career,

_____

_____

_____ !

Chapter 2

Dream

Life seldom presents a "do over"!

Following military service, the question arises, "What kind of career should I pursue?" Reflecting back on my transition, this was a puzzling question that I was not equipped to answer.

Consider military service and the system of career management. This is a unique system in that job placement is done based on quantifiable aptitude and abilities and the needs of the armed forces. You may want to be an intelligence analyst, but if your ASVAB scores don't support that, that field is not offered. A great quality of the military talent management system is its prescriptive nature. (I say that in a very positive way). A service member will be told their job, unit of assignment, and location. What a wonderful system of clarity.

This path continues throughout the military career. Schools are intermixed with other professional development opportunities, but a service member will always have a guiding hand for career direction and advice.

Fast forward to the time of transition, one's military career is ending and now the service member who has had a career trajectory and a guiding hand, a branch manager or assignment officer, is now on their own to figure out what is next. We ask a person who has had limited exposure to the world outside the armed forces to now be able to review companies, career fields, and job postings and make a decision for their next career.

Veterans attract veterans. Without any preparation, a transitioning service member will look at those colleagues who have already left and follow in their footsteps.

Veterans often are the largest limiting factor in their post service career search. One will look at others who have transitioned and listen to the employment mantra of: "You should apply your military skills and work in security or work for a defense contractor." These are wonderful career fields but by no means are they the only options.

As a transitioning service member, you have the opportunity to reinvent yourself. You can vastly change career fields, shake off the DOD employment cycle, and step out and pursue that "dream" job or company. If someone tells you that all you are qualified for is security, they likely have no exposure to military service and are simply repeating tired talking points.

What Are You Capable Of?

In Chapter One, you learned to be proud of your service and your experience. Now, how do you break that down to truly understand what value you bring to a new employer so you can dream about the possible careers you could have?

Your Value Formula

As you assess where you are today and where you must be as you begin to put your name out into the next employment world, everything you do and will present comes down to a simple formula:

TDR + KSA = Your VALUE to be marketed

1. Every role, position, and job description details simply the tasks, duties, and responsibilities **(TDRs)** that it entails and that one has had the opportunity to demonstrate an understanding and level of execution within.

2. Every role, position, and job description details an associated list of knowledge, skills, and abilities **(KSAs)** that correspond to each TDR that one should have.

As you reflect on your service, become aware of what language the corresponding civilian world uses. Don't omit the military language, rather wordsmith yourself into both worlds. What you put on your resume must tell a story about who you are and the value you can bring to others, in this case a civilian employer at either the senior level, mid-management or employee level. Recognize at what level you operate(d) within the military and where that corresponds in the civilian world.

Think of any organization as a simple pyramid diagram, no matter how large, global or lean and local, it all plays out on their levels and your ability to build your brand, presence, and credentials. When others can see where you fit, they will fast-track your transition. The problem most active duty military, veterans, and private sector employers struggle with is the understanding of alignment between the two worlds. Luckily, it is not that hard.

Whether you plan to take the typical transition pathway and look for your entry point with a civilian contractor to the federal government or you want to explore a true private sector opportunity, begin thinking against a calendar as early as possible. Ask yourself:

1. What is my end date?

2. What maximum schooling do I need to accomplish, and in what chronological order?

3. What experiences and position titles must I perfect?

What Do You Bring to the Table?

In Chapter 7, we will discuss how to develop yourself as a brand. This brand represents how you present yourself to the world. However, before you can create this brand, you need to determine what value you can bring to a future employer.

I (Dr. Jeff) have created the Player Capability Index Model™ which spells out how to develop TALENT from an employer's perspective and assess the viability of an individual to predict future success – in a moment, we will apply this to you!

First, know that entrepreneurial spirit and energy are at the root of any success in America – and the global pandemic of 2020 served as a devastating reboot on the human psyche. USA Today and Gallup research reveal that as much as 71% of society at any given time is looking to perform the minimal work possible while expecting the maximum pay. It is no wonder that the entrepreneur has become the rarest species!

Adding to the new reboot and noise is the new reality influencing both the personal and professional landscape. Today it's as if anyone with a mouth assumes they should be a podcaster and everyone with a finger believes they are a blogger. And, everyone that is unemployed or has only held a job for two years at a time feels they are a coach, consultant, or trainer – the NOISE is deafening and distracting.

As a military professional you are well positioned to blow through this noise as a viable alternative.

Understanding the difference between the mind of the complacent and the incompetent and the mind of the successful professional is critical to self-acceleration and identifying within others that rare talent gene called "entrepreneurial."

These individuals act as catalysts to wonderment, innovation (not imitation), advancement, and capital market generation. They have a methodical mental DNA blueprint for creation and a GPS for advancement. Many also understand their lack of business acumen and therefore are continually looking for business management talent acquisition. The mental DNA has a mystical axis, weighted disproportionality as an entrepreneurial AmeriCAN instead of the USA TODAY/Gallup research of the newly shaped AmeriCANT.

The Entrepreneur's Mind

So, what are some of the mental DNA characteristics of an entrepreneur's mind? Well, they are, in fact, CRAZY. They have:

- **C**reative approaches to the obvious which reveal alternate pathways to and beyond a goal.

- **R**esults-oriented responsiveness to market needs and demands and foresight to needs yet revealed.

- **A**ttitudes of victory and self-belief that radiates out and becomes contagious, drawing others in to help them achieve their goals.

- **Z**est for the unknown and a belief in possibilities and a sense of urgency to leverage, capitalize, and execute ROI. They find connectivity where others do not and continuously explore application opportunities to learn, apply, and advance.

- **Y**earning to see the best in others and situations to consistently improve life. You will find that the solo entrepreneurs surround themselves with like-minded challengers and advocates, fueling their sense of CRAZY (as viewed through the lens of outsiders). Yet, they manage to balance productive ROI for right now and evolving ROI for tomorrow.

The Player Capability Index™

To further understand the CRAZY mind of the solo entrepreneur, we must turn to the "Player Capability Index™" as the mental architecture for life-long development. Over the past two decades, I've worked with clients ranging from NASA and the DOD to Harley-Davidson and Farm Credit Services of America banking groups to the National Guard and NASBA. My experience has taught me that understanding the human capital talent within an entrepreneur or institutional employee comes down to a simple matrix I call "The Player Capability Index Model™."

The more you can objectively understand the depth of what each "letter" represents in yourself and others, the more you will realize the entrepreneurial energy and the capacity of any person you meet. This formula can drive 360 degrees of talent modeling (job scoping, hiring, interviewing, task management, promotion, coaching, identifying client needs, etc.). The letters reveal both where the individual is currently and what must be done to enable them to succeed moving forward.

Here is your formula in managing your career and transition process.

$$C = (T^2 + A + P + E + C) \, E^2 \times R^2 = R$$

R = Results

The last letter in the formula represents results, any output or ROI desired.

C = Capability

Capability is the driver of the solo entrepreneur that enables significant results to be continuously generated. The greater the depth of any and every subsequent letter enables the results. Conversely, the complacent among us do not draw upon any lettered capability driver, nor do they add any real-time relevant depth to any lettered category. This serves as the cancer of entrepreneurialism.

The letters within the parenthesis drive the capability level. They are:

T2 = Training

Training is any deliverable of knowledge, whether formal or informal education, technical or non-technical education, certification-driven, or simply OTJ training. The number two represents two applications of the T. The first T is for total training gained from birth to the present. The second T is any training they will do in the future. Entrepreneurs are always seeking more training!

A = Attitude

Attitude must project a winner and not a whiner. Exhibited as one's passion, dedication, work ethic, commitment to execution ... Review your social media footprint as an example and ask, what does this project to others that do not know me?

P = Performance

Performance reflective of past accomplishments serves to bolster self-belief and create awareness and confidence of what can be done in the future ... If other high achievers around you were asked to tell someone that does not know you about you, what would they say about you? What does your quantifiable track record really say ...?

E = Experiences

Experiences from birth to present tense are enormous windows through which entrepreneurs see themselves and from which one can draw strategically for entrepreneurial results ... What have you done and where have you been, and how can that be leveraged to the benefit of others?

C = Culture

Cultural awareness and upbringing also calibrate performance and self-worth. You can draw upon this to know what you can manage. Conversely, what you fail to recognize in your past could spell disaster ... What do you bring as situational awareness and background to gain better understandings of others and ways to align and serve in a forward manner?

E2 = Expectations

Expectations calibrate what really shows up, the entrepreneur or the complacent individual. The first interpretation of E is your expectations of yourself. How you see yourself calibrates whether you bring your A-game or B-game to the show. The second E is the other person's expectations of you. Knowing the two and calibrating them together allows for entrepreneurial effectiveness.

R2 = Relationships

Relationships serve as the multiplier to the entire formula. Knowing how to leverage these relationships is the secret to entrepreneurial success. Unfortunately, far too many entrepreneurs today have their energy snuffed out because of the ever-increasing circle of negative influencers and stimulants around them. Be careful with whom you spend your time!

There are two sets of relationship imprints upon an individual: the first is representative of "past tense" relationships and the second interpretation of relationships is "future tense," those future relationships that can be strategically and purposefully forged that can influence your successes.

Understanding the solo entrepreneur mind and how to take control of your destiny by consistently using the "Player Capability Index" is the DNA that CRAZY is made from!

This model is a critical tool developed in our Dr. Jeffrey Magee | Leadership Mastery program used at graduate schools, Fortune 500 L&D centers, in our leadership development programs globally, and is a key element in our graduate management text series *THE MANAGERIAL LEADERSHIP BIBLE* (PEARSON EDUCATION) and our personal development book *YOUR TRAJECTORY CODE* (WILEY)_____ https://www.jeffreymagee.com/books.cfm.

Entrepreneurial spirit and energy are at the root of any success – and lifelong learners are lifelong achievers. And, achievers are always learning. You can only replicate and steal others' identities, intellectual property, and resumes so long, eventually your BRAND is either the real deal or a mere imitation – who are YOU and who are those you lead, develop, ASSOCIATE with, or tolerate?

Use this Player Capability Index model to craft your present military career for success and use it to guide you in your career transition. For example, let the "R" for results, which is the output to the formula, serve as an industry you want to transition into or if you know of an exact business that you would like to transition into. What would be the ideal position, job, or role within that of interest to you? Let that be a specific "R." To be a viable candidate for hiring, what would that role require or ideally like to see a candidate for interview and hire possess, as it comes to each of the letters within the model – the "T" or "A" or "P" or "E" or "C" that would place you at the top of the consideration of candidates applying?

Remember, the greater the depth of content you have for each letter in the formula, the greater value you can demonstrate to prospective employers that you can bring to them. From this depth of content under each letter, you can explore what actual line items rise to the top for differing positions, individuals, and organizations you may be interested in after you transition away from your military career. The depth, your maturity, your failures, your wins, and your accomplishments have value to a prospective employer. Your network may even have value to a prospective employer – so ask yourself what value do you represent, and then we can explore ways to monetize that value in your next career!

Just as the "R" for relationships can also aid you in your transition, by both evaluating your existing contacts as "Relationships" that can aid your endeavors and when that draws a blank, then look at the "R" for relationships as a guide to the types of people or actual individuals that you need to be getting connected to and or introducing yourself to. We will discuss more on this in later chapters.

Now that you have an objective matrix to use in evaluating who you are and what value points you have to take with you as you transition, let's look at a way to track your progress.

Dream About Your Future

Dreaming about one's future assignments while serving is not a common practice. A prescriptive system for career assignments does not support imagination around what your next job will be. Telling a service member to "dream" about your fantasy or perfect job is asking them to do something where there is no experience.

Let's break this down into tangible bites. Before you can dream about a future career and or company, you need to know what it is you enjoy. For this discussion, let's assume you enjoyed your time of military service. Over a career, you had responsibilities, jobs, or additional duties. Some of them really inspired you, some you tolerated, and others you would not want to repeat. Now, look across your time of service by position and list the tasks, duties, and responsibilities (TDRs) you held, experienced, gained, and learned. This is where the value you represent comes from when others look at you from your overall time in the military. Do not bracket this by assignment, rather just pull from your assignments those activities you really enjoyed. And always include the knowledge, skills, and abilities (KSAs) you acquired and possess.

At the end of this exercise, you will have a thorough list of job responsibilities that brought you tremendous joy. A few items on my list include the following:

- Public speaking or facilitating
- Helping or guiding others towards improvement
- Teaching
- Applying metrics and measurements for decision making
- Leading and coaching
- Teaching financial literacy
- Being around or working with people of high moral character
- Providing recognition to others for exceptional work

I then took my list of items that bring me joy and coupled them with my guiding thesis:

1. I wanted to depart the military on my terms and leave feeling positive.
2. I did not want to work as a contractor or state or federal employee. (There is nothing wrong with those jobs, they were just not for me.)
3. I wanted to work in a vastly different industry than defense.

I now have a framework from which I can start to review industries, companies, and specific jobs.

Step Out of Your Comfort Zone

It can be tempting to follow in the footsteps of other servicemen who have transitioned out of the military. It can feel "safe" to take the advice and stick to adjacent industries like security.

Take a moment and think about your time in the military. You are a member of the world's best fighting force. You have completed courses in leadership taught by the finest leadership institution of this country. You have performed your job in austere and ambiguous environments and made good decisions with incomplete information. You are in a field that most of the country is not qualified for; you have global experience, are physically fit, morally sound, resilient, and live by a set of values. You are a ROCK STAR!! You don't play it safe. You play it full out.

Given this long list of qualities, attributes, and experience, why would you choose to play it safe and only seek out "comfortable" employment?

Now is the time for you to be a bit uncomfortable. You have nothing to lose by stepping away from your DOD/military industrial complex, especially, if you start your transition 2-5 years out. With time as your ally, you can fail and repeat numerous times. You can explore industries, career fields, and build your network, and all the time you are constantly learning and focusing on specific jobs and companies.

Now's the time to step out of that comfort zone, forge a new path, and follow your passion.

Military service has a label for the veteran and the employer considering the veteran. Often, qualifications or attributes revolve around the branch, the rank, and assignments. This can be an extremely narrow view and discounts an endless array of experiences and attributes.

If I focus on branch, rank, and assignment, my offering would be:

- Branch: Army
- Rank: Colonel
- Assignment: G1 (Chief of Staff Personnel)

Think about someone with a combat arms job skill:

- Branch: Army
- Rack: Staff Sergeant
- Assignment: Tank Commander

To relate what you have done in the military to a civilian position, focus on the attributes of the military assignments rather than the rank or titles. If you apply a "standard" combat arms definition of service as the centerpiece of a combined arms that is chartered with closing and destroying the enemy, your ability to step away from your comfort will be very limited. All opportunities will be viewed through the lens of a combat arms professional.

For a tank commander (Staff Sergeant), I would highlight:

- * Leader of a technology-based team with systems totaling more than $10M
- * Decision Making: skilled at making rapid decisions with a broad team focus and done with imperfect information
- * Coordination: member of a multi-discipline team chartered with timely coordination across the dimensions of air and land
- * Leader of the world's premier system, chartered with being the centerpiece of any operation

- * Responsible for the life, welfare, and education of a team of 4-6 individuals
- * Leadership: graduated from the nation's premier educational institute and exercised these skills in numerous multi-national environments

Note, when a combat arms position description is recrafted to focus on the skills it requires to accomplish the job, the opportunity for numerous positions opens. From the few noted items, positions in technology operations, crisis management, leadership, business development, maintenance management, and many others become possible.

This skill of shifting your competencies from rank and MOS to skills and attributes (KSAs) applies to shifting and broadening your job focus. This shift will permit you to step away from what is comfortable. It opens the opportunities for veterans with a combat-focused job skill to step away from a traditional security or law enforcement role and embrace whatever interests them.

You will need to adopt this practice when applying for jobs as well. We will discuss this in more detail in later chapters, but keep in mind that when an employer sees the job title of Tank Commander, their immediate emotional response might be: what can a commander of a tank do in my business? "My last check, my company is not rolling across the Great Plains with the intent of conquering and taking territory." Repositioning and rebranding a military job title will help a potential employer understand the value you bring.

Stepping away from comfort is a challenge for anyone, and for those who served our nation, it is stepping away from a "lifestyle," not just a job. Having been a member of the world's premier armed forces, your offerings are near endless. Your offerings around education, job skills, values, resilience, and leadership are nearly endless. Don't limit your future employment opportunities based on your rank and military job title. Break that mold and see the broad array of possibilities that are in front of you!

Transitioning from the military to the next chapter of your life provides a "Do Over" opportunity. Be strategic and make this new career the dream you have always seen for yourself. Whether you are making your transition after a life in the military or you are on a shorter timeline after your last "enlistment" and ready to transition, the military has provided you massive opportunities for relationship building, education gain, on-the-job experience, accomplishment, and leadership.

Call-to-Action:

Now reflect on what you have just read, ideas that have been activated in your head, and weigh them against each element of the **VALOR™ Model** and what you need to be doing right now at this moment in your career/professional life and with what makes sense for you ...

<u>VALOR: The Ultimate Handbook on Military Transition</u>

What are you doing and need to be doing with respect to the strategies and tactics to raise your **VISIBILITY**,

_____

_____

_____

Illustrate ways for you to gain **AWARENESS** into the market-place,
How can you **LEVERAGE** yourself for your next **OPPOR-TUNITY**,

_____

_____

_____

While being strategic in the **RELATIONSHIPS** you have now, who are they,

_____

_____

_____

_____

And who can you identify that you don't know and need to get connected to for potential advocacy in your later career,

_____

_____

_____ !

Chapter 3

Plan

In the last two chapters, the discussion revolved around accepting the closure of a military career so you can start dreaming of the possibilities. Now, it's time to begin the planning process for a targeted, well-resourced transition. It is imperative to emphasize (if I were teaching a military class, I would stomp my foot on the floor) you cannot move to a planning phase until you fully accept and embrace that your military career is coming to a close!

While you may (or may not) have had the exact military trajectory experience you wanted, you needn't spend too much time in the rear-view mirror. Now is the time to start embracing what you have accomplished and focusing on the windshield of your future. The greater your visibility is in front of you and the further out over the horizon you can see, the greater your options and opportunities will be.

Early during my five-year journey, my wife asked me, "Why are you spending time planning your post-military career? You're not going to leave the Army today." My response was, "If I do nothing, that is exactly what will happen- nothing."

<u>Be as planful with your military transition as you were for a military operation.</u>

The goal of any transitioning service member is to apply the same rigor and detailed planning to your transition to new employment as you would for any military operation. You are now writing an operations order, and the operation is YOU. Apply the identical methods of backwards planning, milestones, communication, and coordination as you would for a military operation. When you served, you would never just walk into an operation the day of with an "I'll just figure this out as I go along" philosophy. This is a surefire recipe for failure. You have exceptional planning skills, skills that have been honed over years of service. Do not let these skills sit on the shelf when you are planning the future for you and your family.

A proven method of learning is to teach from a known point of knowledge; with this, one is moving a student a small degree as the instructor is building onto existing knowledge. Let's incorporate your existing knowledge, develop an operations order, and apply it to your transition.

Operations Order

A common planning tool is an Operations Order. The Army utilizes a 5-paragraph format: **Situation, Mission, Execution, Administration & Logistics and Command & Signal.** With this, you now have an outline or template to affect and coordinate the execution of your transition operation. The following is an example:

Situation: After 8 years of military service, I have elected to transition and pursue civilian employment in the field of technology. I will build my brand around the technical skills acquired from my service, an exceptional education and practice in the art and science of leadership, and my service in the world's preeminent fighting force.

Mission: Over the next 24 months, I will incorporate a self-assessment and gap analysis targeted toward employment in the technology sector coupled with a review of the many resources available to me as a transitioning veteran. I will seek an employer that has a similar set of values as the armed forces and will strive to volunteer with a veteran-serving nonprofit as a means of giving back to my community.

Administration & Logistics:

Over the next 24 months, the following milestones will be met:

- Review my military career with a focus on what brought me happiness and joy
- Align the previous step with fields of employment
- Review job postings and compare my qualifications
- Take steps to mitigate any educational gaps
- Build my industry relationships
- Build my industry language

Command & Signal:

Military purpose: Command identifies the chain of command and their location before, during, and after the operations. Signal provides signal instructions, call signs, frequencies...

- Build and document a network of industry relationships to include contact information
- Affiliate with industry associations (if you're an HR person join the Society for Human Resource Management (SHRM))
- Build a communication plan with a focus on outreach and outcomes on engaging with your network

My transition plan started 5 years prior to my eventual departure. It was not well crafted; in fact, it was not much more than a timeline with a few milestones. What I did have was plenty of time. Time was an ally and not my enemy. With time, I was able to:

- Establish my North Star for post-military employment
- Review various industries and locations
- Build a vast network of business professionals, educators, and colleagues who already transitioned
- Upskill my education with a second master's degree (MBA)

Establishing a North Star to Guide My Actions for Military Transition

My initial task in planning my transition was crafting broad themes that would guide me toward specific actions.

Broad themes:

1. I wanted to depart the military on my terms and leave feeling positive.
2. I did not want to work as a contractor or state or federal employee. (There is nothing wrong with those jobs, they were just not for me).
3. I wanted to work in a vastly different industry than defense.

Let's break down how the above three themes drove my planning and behavior.

1. I wanted to depart the military on my terms and leave feeling positive.

As a commissioned officer, I had 28 years of commissioned service up to the rank of Lieutenant Colonel. I knew exactly when that date was in the future, and I used that as my point from which I would backwards plan my transition. My goal, in alignment with theme number one, was not to hit that date and be told I have to leave. I saw many Soldiers who ran their career up to their mandatory removal date, had no plan for future employment, and left the organization bitter and angry. I wanted to be fully prepared prior to hitting my 28 years. As it turned out, I was promoted to the rank of Colonel and my timeline increased a bit. Despite this promotion, I stayed focused on my timeline and broad themes.

To depart the Army on my terms, three tasks are implied. First, I had to be able to secure employment outside of a contractor or the state or federal government prior to my mandatory removal date. Second, I had to figure out what industry I wanted to work in, and third, I had to figure out how to get a job.

I want to circle back to time for a moment. Having little experience in job acquisition, I wanted plenty of time to figure this out. With ample time, I could make mistakes that would have no cost and serve as a learning experience. Starting this journey early made my eventual departure from the Army easier. I was easing into and learning small bits of information over an extended period of time, and it made me feel that there was a positive outcome somewhere in the future.

I did not want to work as a contractor or state or federal employee. (There is nothing wrong with those jobs, they were just not for me).

Having spent almost three decades in uniform I wanted something different. I did not really know what that different was, but I did not want to return to the Army as a contractor and be that guy who used to be a Colonel. I really wanted a clear break from the Army and a career that offered something new and very different. This statement provided a great deal of clarity and direction. It gave me boundaries and lanes for where to apply my time and focus and, more importantly, where not to apply my time.

I wanted to work in a vastly different industry than defense.

This statement may come off as I want to abandon the culture of military service, a culture that I had thrived in for almost three decades. That was not the case at all. In fact, what I was seeking was a combination of where I came from and where I wanted to go. My military career was wonderful! I have no regrets, in fact if I could do it over again, I would not change one thing. This statement, one that looks simple, is in fact very complicated.

A better way to phrase it would be "I want to work in a vastly different industry than DOD; however, I want the structure, parts of the culture, the camaraderie, and focus on serving a higher purpose all nested in a values-based organization." Essentially, many of the elements of military service without being in military service.

Tracking Your Progress

In my book, *YOUR TRAJECTORY CODE,* I outline a model you can use to mentally and physically track your actions through this career transition. This can aid you in your career makeover.

Goals are essential to every individual and organization. However, setting those goals is meaningless unless you pay attention to the course you're traveling. We all start at Point A en route to our goals, objectives and ultimately success (Point C). Unfortunately, far too often, individuals and organizations get off track and get lost in dead-end behaviors, derailment, and failures (Point B).

You must be deliberate and purposeful as you work towards your goals. Measure your progress along the way with objectives and specific Key Performance Indicators (KPI) and milestones. These benchmarks, and ultimately your final goal, are created from your values which drive your inward vision and outwardly your own mission or purpose statements (MAPS: Mental Action Plans).

To reach your goals and attain a higher level of ROI (Return on Investment) in every endeavor, you must first understand the Trajectory Code Model™.

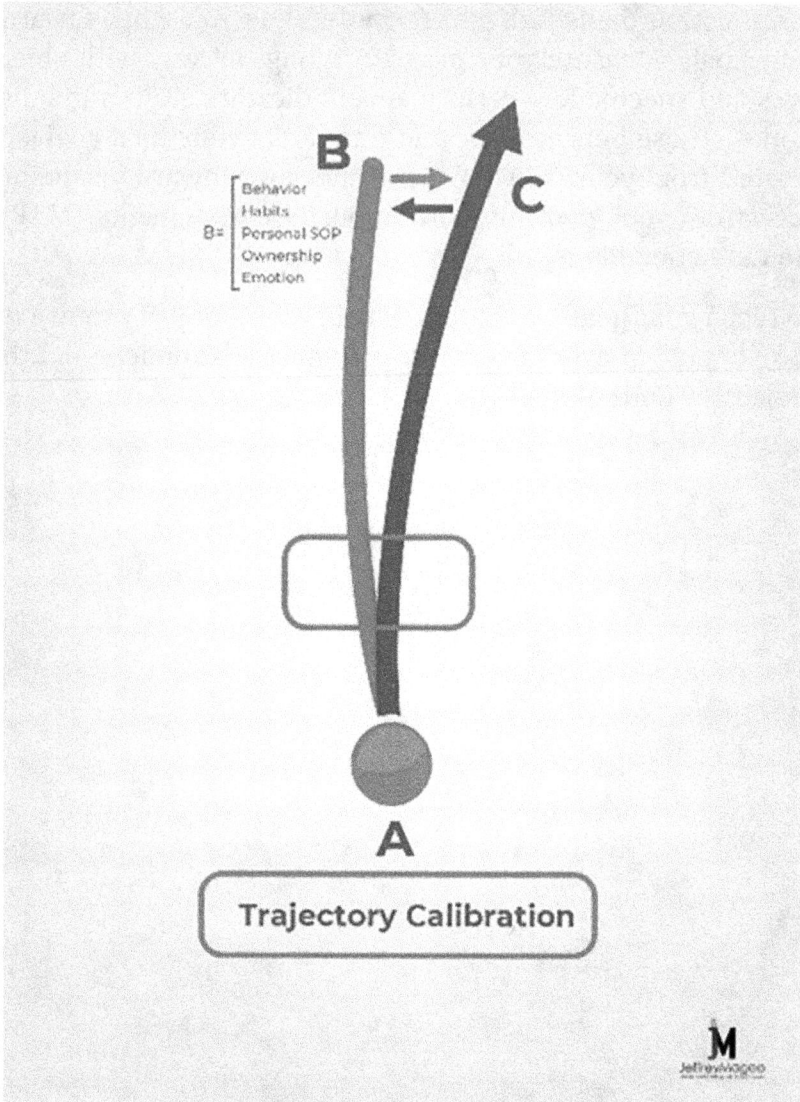

Let me explain this V-shaped diagram. As you leave Point A (the starting point), there is a short period of time where your actions and behaviors place you on a trajectory toward success (Point C) or failure (Point B).

In reality, most people and organizations have the best intentions when leaving Point A (the onboarding of a new employee, starting a new campaign or initiative, etc.). However, unless there are mindful individuals, coaches, mentors, benchmarks, or systems to evaluate progress continuously, your supposed trajectory towards C can send you towards B.

Think of the KPIs that keep one en route to Point C goals as the GPS for accelerated growth. With constant progress evaluations, it's relatively simple to recalibrate and get back on track if you find yourself off course. As a result, success will be more easily (and often) attained.

However, if you head off course with no accountability mechanisms or accountability partner in place, your incorrect behaviors will lead you directly to Point B... far away from your true goal.

Over time, this behavior becomes habit-forming, and habits become your personal SOP (Standard Operating Procedures). At this level, it becomes ingrained to operate and see things only from this dead-end trajectory. This becomes your vested emotions, and your emotions cause you to insist that your way (SOP) is the only way and the best way.

If you reach the dead-end of Point B, the explosion causes defensive behavior, blame, and gamesmanship. The necessary change to get from Point B to Point C seems far too overwhelming for most at this point. In a career transition, this reveals itself when someone has not planned and leveraged a transition plan that will allow them to be successful and chooses to blame everyone else for their career transition implosion.

Imagine if, at the base of the Trajectory Model at Point A, we had people, systems, and tools to help us make the recalibrations necessary to arrive at Point C? I call these easy adjustments the

1% factors. There are endless intersecting and progressing ROIs that can be plotted into the Pathway C Trajectory in the early stages and will get you where you need to go.

When you operate from a greater understanding of the Trajectory Diagram, you can manage and predict human behavior more accurately and facilitate simple 1% change calibrations to attain greatness. You can set benchmarks along the A-C Trajectory to calibrate and re-calibrate when necessary your every action to accelerate to your success. For example, consider the new ROI vocabulary that your Trajectory Code can drive:

1. ***ROIntellect*** –Demand that you continuously enhance your mental DNA and draw deeply from within, always showcasing the best of you and others!

   Your knowledge gained through certifications, licenses, and degrees adds to this ROI. More on-the-job (OTJ) experience and exposure to new undertakings also expands this ROI!

2. ***ROIndividual Initiative*** – Imagine that your Trajectory "C" line was paved with support systems and people to allow you to deliver on ROI #1, so you and others freely gave 100%, 100% of the time and accepted nothing less from everyone in your space?

   The more you have achieved with others and on your own allows for you to have more KPI stacked up here, for inclusion on your resume, into your LinkedIn Profile, and to use in conversation with others and in interviews as ways to significantly differentiate you from others.

3. ***ROInterpersonal Relationships*** (leverage multiplier) – Now, imagine you could leverage 100% of the people you know and could motivate others to do the same in pursuit of Trajectory Point C.

Here is the time to take stock in the years of interactions with peers, subordinates, and senior NCO and Officer leaders around you. Who have you impressed, who has singled you out with citations, awards, recognitions, etc. and have you maintained those relationships or not? Think of all of the volunteer and professional organizations/groups aside from just your job that you have been associated with and/or need to start being engaged with as appropriate for your age, tenure, rank, interest, and potential transition goals.

When you have clarity of Point C with clearly defined language that guides every action, it becomes a common DNA thread that unites people of like cause, mind, or goal with one another, and the cumulative energies become overwhelming. This spells real ROI of any capital you get to work within!

What the marketplace of tomorrow really needs are people who don't just work, apply, contribute, and assist in times of change (the minimum bar of entry to the workplace). What is really desired is the person that can go to the next level to make an impact, a profound difference - someone who can be transformational in all that they do.

Call-to-Action:

Now reflect on what you have just read, ideas that have been activated in your head and weigh them against each element of the **VALOR™ Model** and what you need to be doing right now, at this moment in your career/professional life and with what makes sense for you ...

VALOR: The Ultimate Handbook on Military Transition

What are you doing and need to be doing in respect to the strategies and tactics to raise your **VISIBILITY**,

_____

_____

_____

Illustrate ways for you to gain **AWARENESS** into the marketplace,
How can you **LEVERAGE** yourself for your next **OPPORTUNITY**,

_____

_____

_____

While being strategic in the **RELATIONSHIPS** you have now, who are they,

_____

_____

_____

_____

And who can you identify that you don't know and need to get connected to for potential advocacy in your later career,

_____

_____

_____ !

Chapter 4

Focus

Focus is the imperative here!

As you look towards your military career transition, focus is critical. You will have far too much information coming at you, far too many people willing to give you advice, and far too many possible job options in front of you. You must get and stay focused.

Guiding Questions

To help you stay focused and make this transition successful, here are a few questions to consider:

1. What do you really want next? Are you looking for a job because you need income, or is this a second chapter and your next potential career opportunity?

Making ends meet is important. However, most should strive for a career with purpose.

Think about your time in uniform. You had a clear and specific purpose; it was not the least bit ambiguous. You took an oath to support and defend the Constitution of the United States against all enemies, foreign and domestic. Your purpose in your current profession could not be clearer!

Moving to your next phase, understanding a clear purpose for your job and the company is critical. As has been said many times, just because you remove the uniform, that does not mean that your military habits or outlook will change. When I retired, my new employer's purpose was simple and direct: "**To entertain, inform, and inspire people around the globe through the power of unparalleled storytelling.**"

2. Is the area that you are considering entering a fad and will be a short-term player and opportunity or does it answer and deliver a long-term solution to the market as a need?

<u>Stability and Industries with Staying Power:</u>

There are few careers that are more stable than a career with the military. By and large, layoffs, furloughs, and downsizing are simply not done. You have a defined career path, a series of schools you must attend, and a reasonable idea of a timeline for advancement. Corporate life is quite different. In my 10 years post-military, I have experienced a furlough and seen colleagues who were laid off.

To me, stability in employment is very important. Following a long career or service, moving from company to company is simply not appealing. When you are planning your military transition, take stability into consideration. As a general statement, large, established businesses can offer a greater degree of employment stability than smaller or startup companies. Each has advantages, but you need to figure out your risk tolerance.

3. What type of organizational structure fits you best: command-and-control like the military or more fluid, flexible, independent?

    After having spent the bulk of your career being given direct orders, you may feel most comfortable in this environment and seek out a career that provides a similar structure. Or, you may be interested in the exact opposite! The thought of taking orders may no longer appeal to you, and you may want to look for a role that allows you to self-govern.

4. Do you want to travel or settle down in one geographical area and what do you want that geographical area to be?

    Is there a specific part of the country (or world) you prefer? Do you want to stay in one place for a long period of time, or would you like to live in multiple places? Are you willing to travel as part of your role's responsibilities?

5. What are you non-negotiable values and how do they reveal themselves in your daily thoughts and actions? Does the potential job, career, or business you want to start or business you want to work for match up with your values? Alignment will be critical in this career transition.

    Whether you realize it or not, the values of the armed forces are your personal rudder. Values provide clarity and direction and act as a criterion by which you make decisions. As with purpose, the values of your branch will not evaporate when the uniform is removed. Seek out organizations that have similar values to that of the armed forces. You will find comfort and familiarity with your new employer when values are closely aligned.

Leverage Your Military Values as a Unique Piece of Your Brand...

You will not be the only person applying for jobs, and you will need to find a way to stand out from the competition. Luckily, as a veteran, you have a superpower.

Somewhere, during the interview, you will likely be asked:

"What drives you," or,

"What process do you use to make complicated decisions?"

This is an open door for you to present the values of your branch (that you live by) or a combination of several branches. What a unique way to distinguish yourself!

Look into companies before you apply or interview with them. Review their LinkedIn profile, their website, and if publicly traded their annual report. If you already have a champion inside of this organization, look to them to provide critical intel. When you research a specific organization and their key personalities, you will know their values and be able to illustrate how your values and their values align, in synergy, and will provide value from the moment they hire you.

Highlighting your military values also opens the door for you to present different circumstances during your career where these values guided your decision. Few applicants, especially ones who are younger, can state a set of values and fewer can provide examples of how to apply them. We will discuss using stories to answer questions during your interview in Chapter 10. For now, let's look at two examples of using a story to illustrate your core values.

If asked "how do you make complicated decisions," my response would be:

I have lived my time in the Army by six values: loyalty, duty, respect, honor, selfless service, and personal courage. I have spent 30 years studying, applying, and being graded on my application and adherence to these values. My first screening criteria in making a decision is against these values.

I once counseled another soldier on misrepresenting data, nothing illegitimate, but it was not presenting the entire picture to a commander. We discussed honor and candor and referenced that it is our responsibility to always present an unbiased evaluation despite that information being unpopular. A leader cannot make correct and timely decisions with clouded information.

Regardless of the time you served, you will have many examples of the values of your branch and how you have applied them. Articulating this can provide a true distinction between you and other candidates and provides a reason why you are the best candidate.

You may also be asked something like, "Tell me about a time when you were confronted with an ethical dilemma and how did you handle it?"

A former soldier responded with:

While serving, my team had the responsibility of building relationships with the local population and small group leaders. You may have seen the gear we wear; it can be very intimidating. From the body armor to eye protection, helmet, gloves, and much more. We know that the protective gear was hampering building trust; the ones we were speaking with simply could not see our face or body language. We had a policy within the theater to always wear the gear when off the base. I made the decision that I would remove my helmet, dark glasses, and gloves when speaking to community leaders.

I realized that this was contrary to a directive, but I also knew that I could not complete my mission while wearing all of this gear. I looked to my military values, one of which is respect. I did not see that engaging in relationship building and building trust with the local population to be supported by wearing this intimidating gear. With respect to the local population, I removed a few items. I wanted to come across as a person and not a "military machine."

Your values will guide your decisions, whether on the job or looking for the next one.

Hopefully, these questions will help you focus on the task at hand and determine the best course of action as you step into the next chapter of your professional career.

Call-to-Action:

Now reflect on what you have just read, ideas that have been activated in your head, and weigh them against each element of the **VALOR™ Model** and what you need to be doing right now at this moment in your career/professional life and with what makes sense for you ...

VALOR: The Ultimate Handbook on Military Transition

What are you doing and need to be doing with respect to the strategies and tactics to raise your **VISIBILITY**,

_____

_____

_____

Illustrate ways for you to gain **AWARENESS** into the marketplace,
How can you **LEVERAGE** yourself for your next **OPPORTUNITY**,

_____

_____

_____

While being strategic in the **RELATIONSHIPS** you have now, who are they,

_____

_____

_____

_____

And who can you identify that you don't know and need to get connected to for potential advocacy in your later career,

_____

_____

_____ !

Chapter 5

Build Relationships

Relationships Are Not The Same As Your Network!

Networking while serving may have a negative connotation. It can be associated with a "kiss up," or one who leverages relationships for advancement as opposed to the skills. Consequently, a network is viewed as something that is less than ethical and somewhat distasteful.

For a moment, drop the word network and replace it with relationships.

"Relationship" is defined as the way in which two or more concepts, objects, or people are connected.

While serving, you built many relationships, you had command relationships, staff relationships, peer relationships, and many others. These relationships blossomed and became productive when trust was established. You utilized relationships to review ideas, offer critique, or "war game" a concept or operations plan.

Whether you recognize it or not, you are experienced at building networks; it is now time to shift that skill toward building relationships outside of the armed forces with a focus on a new community.

The old adage "It's not what you know, it's who you know" is still true today. Many of your future career opportunities will come from the people you associate with. This is why it's so important to surround yourself with quality professionals.

It is important to identify early on:

1. The networking groups you should be involved with.
2. The professional trade groups to be attending.
3. The people and businesses on-line to be following, connecting, and engaging in dialogue with.

Within the military organization, there are similar affinity groups that you should be involved with already. Explore the strategically appropriate community volunteer groups and get involved so you can showcase your talents to others and learn more knowledge, skills, and abilities through the tasks, duties, and responsibilities that those organizations can provide to you as well.

Look to your personal network and leverage those individuals more. Six degrees of separation is real. Anyone you want to meet is just six people away from you at any time. With the internet and LinkedIn as an example, you may realize that the degree of separation between you and who you want to be connected with is closer to one or two degrees of connection.

And, always remember that you are showcasing yourself to everyone around you, so make sure that your every action, endeavor, comments, and work product are always as professional, classy, and excellent as possible – you may not get a second chance to make that first impression!

Take a look at the contacts list in your phone. The total number of contacts you have is like a network. Now re-examine all of those contacts and evaluate how many really know you and how many do you really know. How many could you reach out to right now with an ASK where it would not seem the least bit awkward? You have a relationship with those individuals. Being able to reach out to someone that you have an established relationship with is always a different outcome than when you reach out to a network contact.

A network is a group or system of interconnected people whereas a relationship is the way in which two or more concepts, objects, or people are connected. While serving, you built networks and relationships, some formal, as in a command relationship, and others that were informal, a shared experience in a unit or a relationship built during a school.

Often, service members and veterans view the act of networking as distasteful or a less than credible way of advancing one's objectives. Some believe that they can stand on their service record and accomplishments and need nothing beyond that. While serving, this philosophy can stand. But when one leaves service, relationships are critical in advancing business initiatives and even gaining employment!

Upon your departure from the military, your objective should be to have built several deep relationships in industries or companies of interest. Now is the time to create a meaningful engagement outreach campaign!

The Value of Relationships in Your Military Transition

The process of getting hired following military service is a competitive one. Under normal times, a single job posting will receive hundreds of applicants. That group will then get whittled down to 10 or 15 individuals who will be invited to participate in several levels of in-person interviews.

From the standpoint of an employer, hiring is risky. The employer is trying to make a hiring decision based on 4 to 6 personal interactions. They are assessing an applicant's skills, cultural fit, transferable experience, education, inter and Intrapersonal skills, and fit within an existing team. Needless to say, this is very challenging. Be patient and realize that the interview process will in fact be a process. Their schedule may dictate multiple interactions with you, assessments or tests, and then another round of interviews and verifications before an offer will ever appear – timing and planning are critical as you can see.

This is where relationships come into play and is why we stress the importance of having plenty of time for your transition. A relationship built over time with a foundation of mutually interesting topics will give you the opportunity to demonstrate your work ethic, willingness to learn and take criticism, humility, and define your brand outside that of a service member.

You have both a blessing and a curse when it comes to your military affiliation. The blessing is your profession is viewed as being ethical, the best at what you do, and often revered by others. The curse comes from society not understanding exactly what you do. The public's impression of those who serve is based on the news and what is seen in movies. Remember, upwards of 95% of society has never touched military service.

Common associations with military service include:
1. Politically conservative
2. Gun owner
3. Robotic response to problems
4. Not creative
5. Autocratic/dictatorial leader

By building relationships, you will demonstrate your character and demeanor, clear up any misconceptions that may be derived from your military service, and exhibit how you will not only fit within this culture, but enhance it.

Take Time to Build Trust

The relationships between you and the company of interest are a window into your future; you can learn with little downside or cost of failure. My relationship provided a long runway upon which I received advice on my resume and education. I met many executives and had a chance to look at many different lines of business; I learned how the company "spoke" to include the industry language. On the return side, they had a chance to observe and get to know me over many years. In my transition, I built about 10 relationships with my future employer. Some were occasional conversations while others, about 2-3, were very routine meetings. I maintained these relationships for about 5 years. Here are a few actions I took:

- I never asked for a job; I was simply seeking knowledge
- I always took advice seriously: one piece of advice was to get an MBA, and I enrolled in Loyola University and completed my MBA
- I accepted criticism graciously and applied it to improvement
- I always set appointments
- I acknowledged moments of celebration and sent cards and "thank you" notes
- I gave back to the relationship

After 5 or 6 years, I had organically established a strong bond of trust. In 2012, I was on site and preparing to run a marathon. During a conversation with one of my relationships, he asked,

"We are getting ready to build a veteran program, do you know anything about hiring veterans?"

That was the moment when all my hard work paid off. I did not know this question would come but I was prepared with an answer. "Yes!," I said. "I built an employment program for the Army that assisted soldiers who were leaving service to find employment."

That statement started my path toward retirement and a new career. All of this occurred through relationships and had I not built those relationships, my path could have been very different. Long term, deep relationships are a springboard from military service to your next career.

Building trust and a relationship takes time. You do not want to be the desperate veteran who is panicking about finding a job so you rush the relationship. You SHOULD NOT meet a person on Monday and then on Tuesday ask them for a referral for a position vacancy. I doubt they will even entertain your request. It comes off as "tone deaf," and you will likely lose the relationship before it starts. During my relationships, I never asked for a job, my purpose was to learn.

The leading voice today on all factors TRUST would be David Horsager and his work and book under the *Trust Edge* branded name. This is worth your pursuit, add this one to your reading lists.

How to Build Healthy Relationships

Now that you understand the importance of strong relationships, how do you develop interactions that feed your happiness and bring you closer to others, thus forging strong, healthy relationships?

Give to the Relationship

When you take, try to give in return. It can be tough as you are the one who will likely be the benefactor, but you can offer assistance. Items I tried to give in return included:

1. Introductions to the military culture
2. Definition of military rank
3. Introductions to other individuals serving in the Army
4. Discussion of how the Army is managing recruiting

The Relationship Cube™ Model

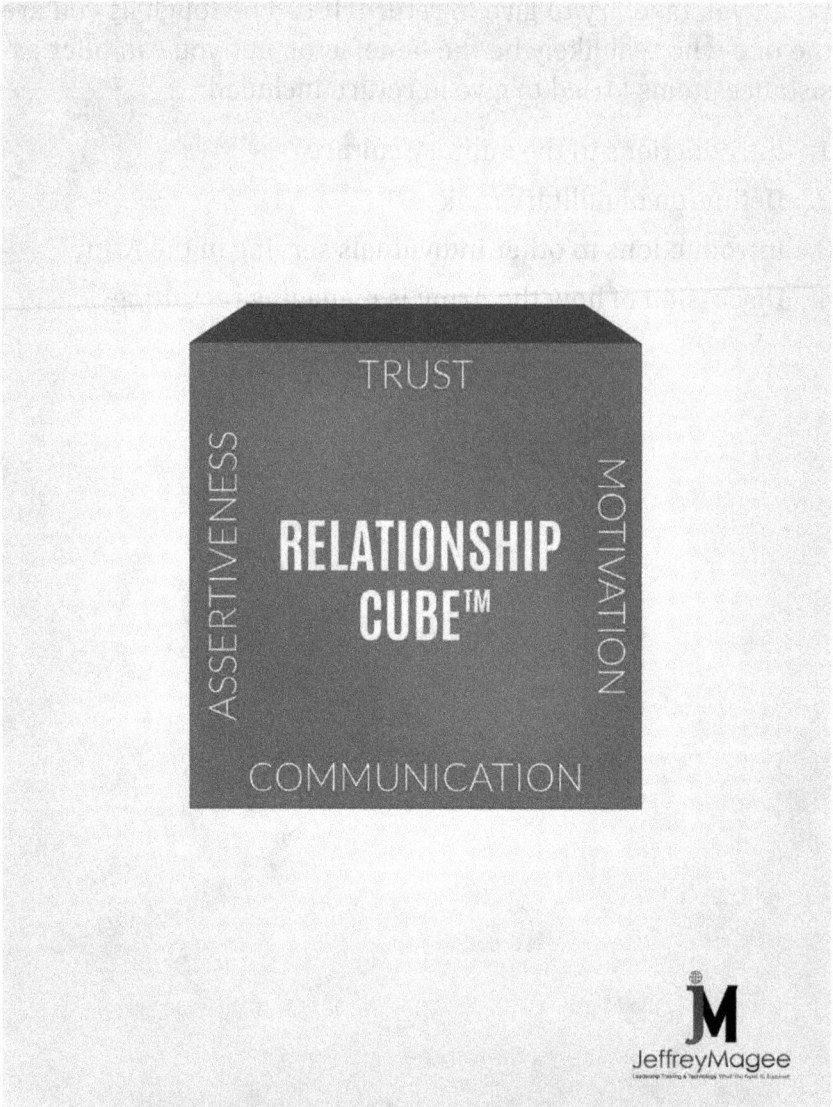

As a Certified Management Consultant (CMC) working with many clients over the years, and speaking with fellow CMC's globally, I have recognized four pathways to happy, engaged, appreciative employees (entry-level/soldier-airman-seaman to the boardroom/senior NCO-officer). This is called the Relationship Cube Model and it can be used to build strong relationships in and out of the office.

To best understand this model, equate it to a banking account. With each person you engage, there is a relationship account between you and them. There are four sides to this account: trust, motivation, communication, and assertiveness, and you can only make deposits into the relationship from one of these four areas. Likewise, we make withdrawals from these accounts from four sides as well. There are endless ways you can do this.

You must understand each person on their terms. What does each side mean to them? How would they like deposits made into it? You'll learn how often you should make deposits and recognize which deposits may actually be counterproductive and rejected. You must always make more deposits into the Relationship Cube™ that you withdraw!

To build and sustain a RELATIONSHIP it is about the deposits first. Below, you'll find a list of universal deposits. Which resonate for you, and are there any you would add?

TRUST Account Deposits
- Honesty
- Do what you say
- Standby others
- Empower others
- Know what others value and garner that
- Keep confidences
- Don't mislead or lie

- Be willing to make the tough calls others shy away from
- Avail oneself to others freely and without the expectation of a deposit in return
- and, more ...

ASSERTIVENESS Account Deposits

- Invite and listen to their opinions and views
- Empower others
- Support others
- Delegate growth tasks and opportunities
- Get out of others' ways
- Help others attain their goals
- Reward appropriately and freely
- Defend them/others
- Compensate appropriately
- and, more ...

MOTIVATION Account Deposits

- Genuine, sincere "thank you"
- Acknowledge successes immediately and personally
- Acknowledge successes immediately and publicly
- Provide meaningful and appropriate awards
- Provide meaningful and appropriate rewards
- Cash, bonuses, pay raises
- Promote without reservation when earned
- Celebrate as others like to celebrate
- Delegate more
- and, more ...

COMMUNICATION Account Deposits

- Listen
- Ask engaging and non-threatening questions
- Make time for engagement
- Communicate to them as they prefer to communicate
- Show sincere interest
- Don't violate things which others value or are offended by
- Be supportive in communication exchanges
- Demonstrate you are willing to take action to communicate with others
- Be conscious of your tone of voice and messaging with others to be respectful
- and, more ...

GENERAL Account Deposits

- You are both immigrants
- You are both first generation American
- You are both first generation military
- You are both from a long family of military veterans
- You are both have commonality among your children
- You are both have commonality among your spouses
- You are both involved in similar hobbies
- You are both from the same geographical area growing up
- You both have shared values
- You both have shared spiritual or religious beliefs
- You both have shared special friends, mentors, people you respect
- You both have had similar deployments

- You have had similar MOS/AFSCs in your military career journey

Understanding which deposits each individual values helps you construct a strong relationship. You will also understand the currency and the currency exchange rate for deposits versus withdrawals with each person.

Respect Their Time and Their Staff

Respect the time of others and assure they are fully present for your conversation by setting appointments rather than expecting them to clear their schedule when you come calling. After you have established a couple of connections, now is the time to start the process of iterative meetings. This is a very delicate line to walk. Walking down the middle, you will be the one who remains top of mind, but you are not a pest. I found that "touch base" calls every 45-90 days was adequate.

Get to know the administrative assistant. You will correspond with them to set up appointments. The term "gate-keeper" is overused but, this person is the one responsible for the time of the executive. I took time to get to know the assistant and after a while, she recognized my area code when I called for appointments.

Appointments are magical! With an appointment, there is a dedicated time slot that is yours. The person you are meeting with can focus their attention on you as opposed to being distracted because of an unscheduled call.

Respect their time. Do not be the one who runs the call past the allocated time. Thirty minutes is a normal meeting time; if they want to run over, that is fine. Most likely, there is another call scheduled immediately after yours. Stage your discussion topics from most to least important. If a meeting is scheduled for

30 minutes, you may only get 15 minutes. Get your important topics out at the start of the meeting. Close the meeting by offering assistance. Remember, you also need to give to the relationship.

When you and those around you are happy, you can deal with challenges more effectively, see the best in others, find satisfaction in work, and strive to do the best job possible. When YOU are happy, everybody wins!

Identify, integrate, and elevate at your present brand level, YOU, Inc., in everything that you do and with everyone you encounter. It is not a bad thing to share strategically with others where you want your military career to take you, and where you want to go in your military career transition.

There are many civilian affinity groups that can accelerate your success post-military. Some of these are:

- YPO – www.YPO.org – Young Presidents' Organization
- VISTAGE – www.vistage.com - Vistage: The World's Largest Executive Coaching Organization
- NAWBO – www.NAWBO.org – National Association of Women Business Owners
- NSA - www.nsaspeaker.org - National Speakers Association

Throughout your military career, as in your transition, it takes work to stay connected. This is not about a once a year cold-call to say hello; it is about knowing them and being an advocate for them, being a PR agent for them, having their best interest at heart, and always helping them to get, grow, and win. So, when it is time for you to brainstorm ideas with them or ask for feedback or an introduction into their network or contacts or even better their relationships, it is natural. The best ways to execute this is face to face, or on video call platforms rather than just a phone call or text.

An Important Note:

Not everyone in your network is a positive influence. Beware of the nomads as they are time suckers, emotion suckers, integrity suckers, and always have an overdrawn relationship account with those around them. They always withdraw from others' accounts and never make meaningful deposits. They are artists at the deception game and leave everyone around them miserable and burned out. These people also will be endless nay-sayers to your every career transition goal.

You may not be able to eliminate all of these people easily from your universe, but you must limit your exposure to them for multiple reasons:

1. Others can view you as them, guilt by association.
2. They can bring you down and feed your skepticism.

Embrace Humility

Here is some advice I received during my period of learning about job acquisition. It was the early days of building my knowledge about how to find a job post-service. I was building a relationship with a recruiter; the focus was learning about how to apply and present my skills, not to apply for a job. We exchanged numerous voicemails before we finally arranged a time to speak on the phone. Yes, this was ages before zoom.

During the call, the recruiter said, "I have to tell you something."

She said, "Your tone on the voicemails is very different from other applicants."

Immediately, I said, "I hope it did not come off as me dictating or ordering."

She said, "Oh no, it just sounded different."

I thanked her for her advice, and we continued to speak.

Here is what I learned from that conversation:

Military communication is unique. It is direct, blunt, and void of many conversational niceties. There is little "chit, chat" during military meetings. We speak in few but impactful words. This style of speech is virtually unconscious.

Had this recruiter not told me this, I would have continued to speak to my external network in my military tone, not even realizing that this can be off-putting. In later chapters, we will address communication and presentation in much more detail. I share this now because it was a time to be humble and accept the advice of a person who was acting as my coach. It paid off.

After I was hired at Disney, one of my long-term relationships approached:

"I really admire the way you took feedback and criticism during this process," she said.

It was by design. I never argued or defended but rather implemented the advice I was given by the people I trusted.

While serving, you are an expert in how the armed forces works. You have years of education, experience, and a vast military network. What you are not an expert in is company culture, processes, procedures, and the value of relationships within a company. Upon leaving the military, you become a novice and must learn an entirely new culture. This will be a test of your humility. In this phase, you will need to place your military ego in neutral.

Now is not the time to be the colonel or master sergeant. Remember, you are now the new and relatively new and inexperienced; position yourself accordingly.

Your network is there to support your transition and provide you with knowledge. Whether it be with mentors or prospective employers, your job is to listen and learn. During my preparation to transition, I took full advantage of anyone who was willing to provide their time and knowledge. This was an opportunity for my learning and development. I also knew that this knowledge, let alone my observations about how companies functioned, would pay off in the future.

"You were always there to learn and receive input and feedback," said one of my mentors. "That is different from many others."

Your Transition Team/Relationships Are Not There to
Inflate Your Ego

Just as everything you accomplish in the military is done through a team, your transition is no different. Leaving service and finding your dream job is entirely new. You are an expert in the military but in this, you are a novice. Keeping that in mind, your transition "board of directors" (which we will discuss in Chapter 6) will help you stay on track and work toward your goal. These individuals will also provide blunt feedback when you start to deviate off your ideal trajectory.

Because of this crucial responsibility, you must be able to accept this feedback with an open mind. It is easy to ask for critical feedback from your trusted inner circle, advisors, and industry experts in your transition process. It's considerably more difficult to actually take it.

My transition team provided invaluable feedback during my journey. When I started my transition, military service and the corporate world were roughly equal in one area: compensation. I am not suggesting they're equal in terms of what is paid across levels, but a dollar is the same dollar be it from DFAS or a corporation. Now I am not any more or less materialistic than the next person but, in this case, compensation made sense across the two institutions.

Keep in mind, I am in the early stages of learning and planning my transition and as a general rule, compensation is not a topic of conversation. Add to that, I did not have the background to properly discuss compensation and in doing so, I was in danger of underselling myself.

One of my mentors, a retired Command Sergeant Major, said: "Don't worry about money, it will find you, you don't need to find it!"

I had been so fixated on money that this proved to be some of the best advice I ever received because it changed my strategy.

1. I stopped worrying about money and title and now focused on work that I would find thrilling and fulfilling.
2. I developed my criteria for what I want an employer to be.
3. I spent my time discussing the purpose of the work as opposed to the salary of the job.

From the perspective of an employer, compensation is a dicey topic. If an employer perceives that one's sole motivation is money, an offer will likely never come. The reason is simple: if money is all that drives you, another company can come along, offer another $5,000, and you will leave.

Employers want people who have passion for the company and their job. What excites you each day is the work you do and where you are doing it. An overemphasis on compensation will detract from your brand and message. As the sergeant major said, "The money will find you."

You Don't Know What You Don't Know

An Army colonel was participating in a fellowship. He and his leader, who was also his mentor, were sitting in a group meeting. A topic was brought up that was specific to the company. The veteran/fellow was about to speak up, but he chose not to. He instead listened to the conversation, and when the meeting was over, he and his boss were reviewing the topics. At this time, the fellow brought up his concerns.

"Boy, I am glad you didn't mention this before," the veteran said. "Here is the company history and why this idea never got off the ground."

Now, you might be thinking: the colonel/fellow is brand new in the company, how is he expected to know the history? He is not expected to know the history but everyone attending the meeting is not aware of his status either (a career army officer who is participating in the Hiring Our Heroes fellowship program as part of his military transition). In that meeting, he is viewed as another member of the team.

Listen more than you speak. You'll be amazed at what you learn.

Advice at This Point Is Free, Take it and Act On it

One of the many advantages of planning your military transition in advance is you have the opportunity to receive a great deal of input from many different perspectives. Over the five years that I planned my transition, I found most of the advice I received to be helpful and on target. Of course, some counsel was disregarded as it did not fit within my strategy, but I still appreciated it. Your responsibility is to build a broad and diverse network and bring in many different opinions. You will sift and sort and determine those that fit your strategy, the ones that don't, hold onto those too as they may help another veteran.

Resource:

U.S. Chamber of Commerce-Hiring Our Heroes Fellowship

https://www.hiringourheroes.org/career-services/fellow-ships/

The HOH fellowship programs are best-in-class workforce development programs that place highly skilled and educated transitioning service members, veterans, military spouses, and military caregivers with employers committed to hiring them.

Call-to-Action:

Now reflect on what you have just read, ideas that have been activated in your head, and weigh them against each element of the **VALOR™ Model** and what you need to be doing right now at this moment in your career/professional life and with what makes sense for you ...

VALOR: The Ultimate Handbook on Military Transition

What are you doing and need to be doing with respect to the strategies and tactics to raise your **VISIBILITY,**

_____

_____

_____

Illustrate ways for you to gain **AWARENESS** into the marketplace,
How can you **LEVERAGE** yourself for your next **OPPORTUNITY,**

_____

_____

_____

While being strategic in the **RELATIONSHIPS** you have now, who are they,

_____

_____

_____

_____

And who can you identify that you don't know and need to get connected to for potential advocacy in your later career,

_____

_____

_____ !

Chapter 6

Create A Board Of Directors
(Aka Advisory BOD):

Just as a business might have a board of directors in place to aid and guide the senior leaders, you can replicate this value on a personal level. Think of this as your personal advisory board or mastermind or what I call your FIST FACTOR™ (as discussed in my book *YOUR TRAJECTORY CODE*). There are 5 critical influencers on who you are and who you will always be and can become.

Each year, individuals spend countless hours and billions of dollars seeking greater success. So, why do some attain high levels of accomplishment while others never seem to get ahead? Why do some accelerate down the AC-Trajectory of the *Trajectory Code Model* we discussed in Chapter Three and understand the application of the individual elements of the *Player Capability Index Model* discussed in Chapter Two?

Every person has the same DNA for success or failure in today's global opportunity. The difference between those who execute a successful military career transition and those who fail is whether one assumes ownership of the *"5 Critical Influencers on WHO You ARE and Always Will Be"* or excuses them away.

For more than three decades, I have worked on the frontline of performance execution and achievement – from global superstars to the person I see in the mirror daily. From serial entrepreneurs and military generals to solo business operators, Fortune 100 leaders to Olympic athletes, musicians, and celebrities to best-selling authors, and world leaders – the answer is always the same.

It does not matter where you come from, what your roots reveal, or where you are going. The *"5 Critical Influencers on WHO You ARE and Always Will Be"* never changes. What does change is how you assume ownership of each, manage them, and elevate them regularly. These are the mental and physical influencers to every experience you have or ever will have, the source(s) of all education and influence, fact or rhetoric, logic or revisionist history, and emotion stimulation orbiting within you now, past tense, and in your future.

Remember the adage –

Garbage in, garbage out!

Actually, this is incorrect ...

Garbage in, garbage stays!

From birth until our last moments on this earth, I call this imprinting your FIST FACTOR™, the *"5 Critical Influencers on WHO You ARE and Always Will Be".*

You are the sum of every person in your life and their impact upon you.

Embracing the FIST FACTOR™

I'm going to walk you through an exercise I've used with my clients for decades across the world. Flex your hand open, palm up. You have five fingers, now stop and reflect. Who are those people from who you:

- Seek mental solace
- Seek mental guidance and advice
- Are influenced as to how you think, feel, believe, and operate

As you visualize a person, add that to one of your fingers and keep the countdown going. Take a minute to reflect and assign one name per finger. This is your FIST FACTOR.

Now, take your fist and slam it into the palm of your opposite hand. Do you feel that power and energy? This is where your power, strength, energy, self-confidence, beliefs (or lack thereof) come from. This is how you take ownership to make immediate and sustained changes in how you think, feel, and behave. This is where you can influence those around you to greatness. Inventory who you have in your head now that shapes your psychology and pathology and then take ownership of their role in your life.

There are *"5 Critical Influencers on WHO You ARE, and Always Will Be,"* and each one has met a differing side of your personality and character; so let's identify each and understand who they are:

1. **FAMILY –** you don't get to select your FAMILY, but your family is the earliest and potentially most long-lasting influencer. Consider your hand inventory from above, were there any FAMILY names that you assigned to a finger or fingers? Do they elevate your game or derail your abilities? Do you embrace them or allow them to hold you back? Are you holding onto toxicity, and do YOU need to let go (if not physically, then at least mentally let go)?

2. **FRIENDS** - you get to select your FRIENDS, and they too can serve as early and potentially long-lasting influencer groups. Consider your hand inventory from above; were there any FRIEND names you assigned to a finger or fingers? Do they elevate your game or derail your abilities? Do you embrace them or allow them to hold you back? Are you holding onto toxicity, and do YOU need to let go?

3. **PROFESSION** – you get to select your PROFESSIONAL key stakeholders, allowing them into your head as early and potentially long-lasting influencer groups. Consider your hand inventory from above; were there any PROFESSIONAL names that you assigned to a finger or fingers? Do they elevate your game or derail your abilities? Do you embrace them or allow them to hold you back? Are you holding onto toxicity, and do YOU need to let go?

4. **SUCCESS** - you get to select the SUCCESS key stakeholders you allow into your head. The SUCCESS category is that individual you personally know that provides you with meaningful insights and KPIs to replicate and attain accelerated achievement and success in your own trajectories. Consider your hand inventory from above; were there any SUCCESS names that you assigned to a finger or fingers? Do they elevate your game or derail your abilities? Do you embrace them or allow them to hold you back? Are you holding onto toxicity, and do YOU need to let go?

5. **UNDERDOG** – you get to select the UNDERDOG key stakeholders you allow into your head. The UNDERDOG is someone you know who finds a way to prevail over adversity instead of complaining and blaming. We can learn from their mindsets and behaviors how specific KPIs for success look. We can model their greatness and be motivated and inspired through them. Consider your hand inventory from above; were there any UNDERDOG names that you assigned to a finger or fingers? Do they elevate your game or derail

your abilities? Do you embrace them or allow them to hold you back? Are you holding onto toxicity, and do YOU need to let go? Many serial success individuals will say this fifth FIST FACTOR influencer is the most powerful for the body armor one needs to sustain life success!

When it comes to allowing people into your life, consider your future associations strategically through a mental and physical health perspective. They will be key influencers on who you will become, so choose wisely.

How many people did you initially inventory in your FIST Factor? Are there multiple names in any one influence category? Only you can determine if that number is acceptable, too many, or not enough. Do you have any influence category empty of names? If so, you are out-of-balance with the *"5 Critical Influencers on WHO You ARE and Always Will Be"*. While there may not be an exact or correct number of names you want in any one influence category, you do want to ensure you have power names in each influence category for balance-of-perspective and better accountability partners.

Now, let's dramatically elevate this FIST FACTOR concept. Think of these five people categories as key influencers, and let's adjust the name *from FIST FACTOR to board of directors or your advisory board*. You need representation in each influence category to have a balanced and accountable internal dialogue or outward discussion on any factor in life, to elevate your game and ensure you're not being influenced disproportionately in an ill-advised direction.

Every aspect of your life has been shaped, influenced, and guided by one or more of these five influencer groups. How you engage them in your future will reinforce your beliefs and actions or challenge them accordingly.

To achieve balance within your FIST Factor (aka board of directors or advisory board), first, ensure you have representation in each of the five influencer categories. Your consciousness, unconscious bias or not, objectivity and desire for critical analysis capabilities and not just self-validation will be directly influenced by the diversity and balance of viable "legitimate" influencers!

Elevate your life trajectory, as I call it, in my best-selling book, *Your Trajectory Code* (www.JeffreyMagee.com). For any next level goal you have, reflect upon your FIST Factor, and will they get you where you want to go? If so, leverage them and their imprints on your psychology. If you need to expand your five critical influencer categories with new people, expand your network.

On the other hand, if you had an unhealthy member of your board of directors, you would fire them. It's time to unleash yourself from those who no longer provide value to you and replace them with conscious contributors to your life!

Build a strong FIST Factor, filled with great relationships, high achievers in your area of focus, and supportive mentors. This will accelerate your greatness, advocate for you, and stand by you as you move forward.

Your board will be your force multiplier!

Using Your Board of Directors

While serving, you were part of a tea. In fact, this is likely a point of comfort. Given the value of a team, their diverse perspectives, and their ability to help you advance your transition toward success, why would you ever go at it alone? My board of directors were pivotal in my transition and ultimately helped me land at my company of interest!

An advisory board of directors is "your" advisory team, a group of trusted colleagues who will hold you accountable and keep you focused on the objective. I like to think of this as primary staff or my personal fire team because each person will bring a unique skill that is focused on your effective transition.

For your board to keep you focused and hold you accountable, there must be a clear set of objectives with a timeline of milestones. This can take the form of a personal strategy statement with a corresponding timeline.

Example of a strategy statement:

By the time I reach the rank of XXXXX, I want to be in a position where I can depart military service on my terms, having completed my undergraduate degree in business, built a network of relationships in the corporate community, and fully built my brand as a transitioning military veteran.

Nested under this statement can be a timeline that establishes your deadlines to support your strategy. With a strategy statement and a timeline, your board now has several ways to hold you accountable. Take the time to publish and distribute these documents and as time evolves, do not be reluctant to update either.

Who Shall Compose Your Board of directors?

In general, you are seeking individuals with a specific set of knowledge as well as networks that they can open to you. These individuals also need to be comfortable with providing candid advice and holding you accountable. There are four types of board members you will want to recruit.

Board Member Number 1: An individual within the company or industry that represents where you want to be employed.

This valued member can provide an endless amount of advice to include:

- Advice on company culture
- Gap analysis regarding your background and credentials
- Introduction to their networks
- An internal advocate

Board Member Number 2: A veteran who has successfully completed their transition.

Their contributions include:

- Their success and failures in transition
- Counsel regarding claiming benefits of transition or retirement
- A veteran's view of corporate culture

Board Member Number 3: Someone from the academic community.

This can be a college professor or an instructor in a certification program. This is a person with whom you can discuss the latest industry trends to include:

- Relevant academic research focused on your field of study
- Business care studies
- Introductions to their network

Board Member Number 4: A cheerleader.

The transition process is not an easy one and you will have many ups as well as downs. Having a person who knows your strategy and can pick you up when things seemingly fall apart is a critical member of your team.

Here is a look at my personal board of directors and how they came to be a part of my success:

Board Member Number 1: A company or industry representative.

For me, my company of interest was ESPN/Walt Disney Company. About 5-6 years before I was to leave the Army, I set out to meet any individual within Disney. At this early point, I had no relationships inside Disney.

My first attempt was calling the employment telephone number.

"Are there any veterans who work at Disney that I can speak to?"

"We don't do that," they responded. "You need to apply for a role."

"I don't want a job, just a conversation."

That attempt went nowhere, but I was committed.

Next, I asked colleagues in the Army if they had any relationships with anyone at Disney. A second lieutenant happened to have been an intern at Walt Disney World, and he connected me to his recruiter. Within 12 months, I was attending a conference in Orlando, Florida. This recruiter then arranged a meeting with the manager of professional recruiting at Walt Disney World. I was sitting in her office, exchanging pleasantries, when another individual approached. The manager I was meeting with said to another Disney leader, "Hey, you need to meet this guy!"

We shook hands and swapped business cards. This person was a leader within the talent acquisition community for the Walt Disney Company. At the time, I had no idea how valuable this relationship would be, but I spent the next 5 years building it. I was advised to get an MBA. I had ample time, so I set out and got my second master's in business administration from Loyola University, Baltimore, Maryland.

Five years later, this leader was the person who told me of the new position that was being created focused on veteran employment. With her guidance, I submitted my application and got the job.

Board Member Number 2: A veteran who has successfully completed their transition.

This individual may serve as a sounding board or emotional support if needed. Serving in the military is unlike most professions. You and your unit/team share a common purpose and vision, you have taken the same oath, you have trained and perhaps deployed together. By virtue of your time in service, you have status (the rank you hold). You are respected internally and by friends and neighbors.

When you separate or retire, this disappears in an instant. To say it is shocking is an understatement. I distinctly recall (this was 11 years ago) I did my final retirement out processing at Fort Meade, Maryland. Still in uniform, I was walking across the parking lot of the commissary and an E-4 approached me. He saluted, I returned, after about 10 feet of separation, I turned around and spoke to him.

"You will be the last person that ever salutes me," I said. After almost 30 years in uniform, that was my final salute.

Why are we discussing this? This change in your life, a loss of friends, comrades, and sometimes status can be shocking and at times troublesome. Having a friend who has already made the journey, one who can share their methods of entering their next phase of life, can often smooth the ripples of change.

The next unknown that a veteran can assist is in the myriad of steps, processing, and gaining the benefits that you are entitled. I have served this role for many veterans and in turn, taken steps for introductions within companies of interest, connections to experts for VA claims and the processing of, and some smaller items like local or community offerings.

Board Member Number 3: Someone from the academic community.

You may think that this is an odd member of your team? Let's take some time to define this. Part of your transition will likely involve some up-skilling. It might be completion of an undergraduate degree, an advanced degree, or a certification. For any of these endeavors, you will be surrounded by like-minded individuals, both students and the instructor. Take time to build relationships. This is a window into the world outside of military service and a great forum to discuss contemporary issues of your area of interest.

Education is a unique time in a person's life, a time when you focus solely on yourself and your needs and desires. Few times in a person's life can you spend time discussing interesting industry topics. One of my transition tasks was completing my MBA, I chose Loyola University in Baltimore, Maryland. This degree was a realization that my knowledge and practices in the Army were not far off and, in many cases, a perfect match for how business is conducted. It was a very affirming program.

Toward the end of the program, I was in a class on technology, the professor made the , "When you finish your degree, you will miss the academic world. You will miss our discussions for there are few environments where you can sit and discuss intriguing topics. "He was right, I do miss these discussions, and they are hard to replicate outside of a learning environment.

Board Member Number 4: A cheerleader.

Let's take a moment and discuss the "cheerleader," also known as your office of morale and welfare. Whether your transition timeline is as long as 5 years or as short as 6 months, this will be a journey. Having a person who is objective and can pick you up or redirect you when you deviate from your strategy is an invaluable team member.

The two cheerleaders on my team were my wife and mother. Each knew my objective and was involved in the many steps I had to take to achieve it. Both served as a sounding board, my wife, Susan, supported me during my MBA, stepping in with tasks as simple as walking our dogs. To maintain your direction and momentum, having a positive voice or one who is not so deep in the process and can offer objective advice will serve you extremely well.

I have found that a board of directors (or network) is a portion of your transition plan that can easily propel you forward or slow you down. Think back to your time in uniform, virtually everything you accomplished was being a member of a team. Your transition is no different. Use your exceptional team building skills, gather the board of directors, and now you become the mission.

When something is unknown to you, ask them. When something just does not seem right, ask them. When something seems too good to be true, ask them.

A Word of Caution

You need to be smart in building your mental board of directors, this advisory board or FIST Factor we have presented. A smart major business in the global economy has a board of directors (BOD) that operates with clarity of governance. To do this, they have one credentialed, vetted, proven person for each major piece of business operation that a business would have. In essence, one BOD member who has a direct line of accountability and responsibility to each executive or C-Suite professional. So, if a business would have a chief financial officer, then there would be a mirror on the board of this position to be that CFO's sounding board, advisor, and accountability stakeholder. If technology is big for a business, then they might have a chief technology or information officer, and there would be a mirror on the BOD.

As you build out your BOD, think of all of the key areas of "BRAND YOU." Reach out to the power personalities that you have built a trusting relationship with and ask them if they would be willing to serve as a confidant to you in this most valuable role.

Share your new resume with them or ask them to look at your LinkedIn profile and provide feedback before you go prime time.

Here are just some of the valuable people already in your network and which you may already have a level of a relationship with as professionals that could serve as your fist attempt at building your BOD; ask them for a one-on-one meeting to seek their advice, counsel, and support:

1. Do you have an accountant/CPA - ask them to serve on your Advisory BOD?
2. Do you have an attorney - ask them to serve on your Advisory BOD?
3. Do you have a realtor - ask them to serve on your Advisory BOD?
4. Do you have an insurance agent - ask them to serve on your Advisory BOD?

5. Do you have a car dealer that you have recently purchased a vehicle(s) from - ask them to serve on your Advisory BOD?

6. Do you have a regular local restaurant - ask the owner/manager to serve on your Advisory BOD?

7. Do you have a private banker where you bank - ask them to serve on your Advisory BOD?

8. How about that dry cleaners you frequent; that owner has a network and an invaluable insight on business, ask them ...

9. Do you really know your neighbor(s) - that may be a pay day of treasure right next door?

The point is, there are people all around you who may bring value to you during your career transition and beyond.

Your BOD will be our talent pool to ask for advice, get feedback on how you present yourself on-line, through your written materials, and credentials. Ask them for their feedback, and especially ask them for feedback if they know the industry or know the business you're working to get connected into. Your BOD may be able to also serve to connect you to critical personalities in your career transition and allow you to fast track through to interviews and possible job offers.

Call-to-Action:

Now reflect on what you have just read, ideas that have been activated in your head, and weigh them against each element of the **VALOR™ Model** and what you need to be doing right now at this moment in your career/professional life and with what makes sense for you ...

VALOR: The Ultimate Handbook on Military Transition

What are you doing and need to be doing with respect to the strategies and tactics to raise your **VISIBILITY**,

_____

_____

_____

Illustrate ways for you to gain **AWARENESS** into the market-place,
How can you **LEVERAGE** yourself for your next **OPPOR-TUNITY**,

_____

_____

_____

While being strategic in the **RELATIONSHIPS** you have now, who are they,

_____

_____

_____

_____

And who can you identify that you don't know and need to get connected to for potential advocacy in your later career,

_____

_____

_____ !

Chapter 7

Create Your Brand:

Applying the word "brand" to a veteran can be a foreign concept. In the past, you've likely only heard the term associated with a product or service. Brand, applied to a product or service, is a concept that distinguishes one product from another and can be easily communicated.

During your service, your uniform, complete with your awards, badges, and rank was your brand. Did you ever have to make an overt statement about the elements of your uniform? Probably not.

By merely looking at a uniform, one has an idea how long that person served, schools attended, accomplishments, deployments, and level of physical fitness. Without saying one word, a brand has been established.

Compare that to life after service. A common statement after the uniform comes off is, "I am just an anonymous person in a suit." Items and articles that were part of the military culture can be odd in the corporate world.

For example, I used to carry the ACU patterned wallet. It was perfectly functional with a compartment for my ID card on the exterior. Recall that this wallet was sealed with Velcro. This is an item that is normal in military circles, however, in a business environment, a camouflage wallet with a loud Velcro seal does not fit. Invest in a black or brown leather wallet.

I had my very first interview while still on active duty. The company was ESPN, the sports broadcast company that is owned by the Walt Disney Company. They flew me to Bristol, Connecticut for a day of interviews.

My first challenge was attire. I could not wear my uniform. I did not want to wear the standard "interview" attire of khaki pants and a blue blazer. I figured black was a rather conservative choice and easy to match shirts and ties, so I went to Nordstrom and bought a black suit. I had no briefcase, but I did have an OD green aviator's helmet bag with patches from various units sewn on it. It practically screamed, "I SERVE IN THE ARMY!" And yes, I had my ACU patterned Velcro wallet.

While serving, your job, rank, uniform, badges, and awards composed your brand. Upon transition, all of that vanishes! It is now up to you to craft and communicate your brand to an audience that may or may not be familiar with your profession of arms.

If we were to meet today, how you present yourself (any online social media presence you have) and how you look are the beginning impressions of Brand You!

Understand the Stereotypes Surrounding Your Service

Stereotypes surround most brands, and veterans are no different. We have discussed in earlier chapters some of these stereotypes. It is worth a quick review.

Stereotypes:

1. Politically Conservative

2. Gun owner

3. Order taker

4. Lacks creativity

5. Type **A** personality

As you construct your brand, these items fall into a "watch for" category. Some are tougher to address than others, for example political affiliation. (I recommend leaving that alone). Creativity, personality type, and merely serving as an order can be easily addressed, and you can address each of these over a period of time when you are building your relationships.

Build upon the positives of military service & manage the negative stereotypes

Your membership with the United States Armed Forces gives you a tremendous advantage in the job market if you apply your experience and background properly.

Employers are looking for:

Stability and loyalty: In today's job market, people are shifting jobs almost on a whim. I am not criticizing this but for an employer, it's tough when teams are consistently changing team members. You bring an element of stability; you have proven this with your military service, be it 4 or 40 years. One of the statements I commonly use is "I have worked for two organizations over my life, the United States Army and the Walt Disney Company."

Leadership: The armed forces is the finest institution to train leaders in the United States. As an organization, they have been around longer and trained more leaders than any other. Unfortunately, veterans seldom illustrate or define themselves as leaders. This is essential to showing how valuable you would be to a new organization.

When establishing yourself as a leader for potential employers, it's important to discuss the military education system: regardless of if you are enlisted, NCO or officer, you have gone through years of structured leadership development. And YES, basic training or basic combat training (Army) is leadership development.

The Army Leadership Model

The Army Leadership Model comprises attributes (character, presence, and intellect) and competencies (leads, develops, and achieves).

This is an Army model. Each branch has a similar model and from this framework, you can now articulate how you operate as a leader.

Incorporate the following (in your own words) into written or verbal communications with potential employers. Take a moment to illustrate each and remember, you are explaining this to a person who has never served and may have a preconception of what military leadership represents.

Given that this position is a leader of people, I think it is important to address my leadership philosophy, education, and practices.

As a (soldier, officer, NCO) in the United States Army, I have been the recipient of years of formal education solely focused on leadership. I want to be clear: this is not a series of ad hoc discussions, but rather academic focused courses taught in a "college like" environment that build and develop leaders for today's complex and rapidly evolving armed forces.

I have adopted and built my leadership philosophy around three tent pole items: "Be, Know, and Do." I then break these down into attributes and competencies. For example, "Be" incorporates character and presence; "Know" focuses on intellect; and "Do" encompasses leadership, achievement, and development.

I want to take a moment and elaborate on some of the characteristics within attributes and competencies sections.

Let's start with "Presence:" bearing, confidence, fitness, and resilience.

This is a system of characteristics or practices that define a leader and inspire confidence for those who are led.

Presence:

- Bearing and Confidence - I will present an attitude and demeanor of faith in my abilities and that of my organization, and I will represent the position for which I am serving in a professional and ethical manner, free vanity or hubris.

- Fitness - I take time to care for myself, both physically and emotionally. A fit person can support others, work long hours, and present as a thoughtful compassionate leader.

- Resilience - I am equipped to handle success and shortcomings with a focus on bouncing back and bringing others with me.

Achieves:

- Gets Results - I am focused on producing results but balanced with development of team members and placing each person in positions from which they can excel.

- Anticipation – I have agility and responsiveness, the ability to shift priorities should a course of action become untenable.

- Feedback- I have the means whereby all parties can consistently improve, discussion of what went well and how to improve for the future.

These are just a few items that can be discussed that can illustrate your acumen and competence as a leader. Combine the above with some "real life" examples during your time of service and you will distinguish yourself.

Be Mindful of Your Online Footprint

Your every public persona must be communicated in a way that people can see the benefits and value you can bring to their way of life, their goals, and their value system. Often, the first place people will go to research you will be on the internet.

Employment search firm recruiters, employment firms, company recruiters, HR leaders, hiring managers, business owners, casual business acquaintances you make at a community event, your FIST Factor, and even your competition will all be to grab their phones when you are not watching and look up your name. What platforms will they find you on?

1. Facebook
2. Instagram
3. Pinterest
4. Twitter/X
5. LinkedIn
6. TikTok
7. Podcast channel host or guest appearances
8. Online blogger
9. YouTube channel host
10. Your own website

11. Etc.

12. Nothing ... nowhere to be found, and what questions would that trigger?

And more critical is not what your account posting may reveal, but what others linked to you may be saying and posting – remember just as it can be "guilt by association" it can also be "success by association"! Ask yourself in this time of transition what presence do you have in social media and which ones are appropriate and which ones may need to be purged!

Be careful in establishing your brand appearance to the town square of public opinion (which includes your online presence). Once you go public with your position on a:

1. Topic
2. Politics
3. Media
4. Religion
5. Lifestyle
6. Diversity
7. Inclusion
8. Music
9. Product
10. Fade or trend
11. Etc.

Then you have branded yourself and some will be neutral on your position, some may love, and some may be put-off by your position; There's nothing wrong with voicing your opinion, just be ready to own your position and what it will mean for your brand.

Create a Strong Social Media Presence

Due to your military culture, this rather simple task can be quite complicated. You first have to get past the belief that a presence on social media will compromise a level of security. While that works well while serving, your new objective is to build a brand and market your brand on a broad platform.

As we have discussed, the same guidelines that apply to communicating your military career to those who have not served apply to your social media presence. Do not forget:

- No military jargon or acronyms
- Create ways so those who have not served can understand your experience
- Have an objective or purpose that is aligned to your career goals
- Do NOT USE a DA photo as your profile picture - invest in a professional headshot
- Align your military experience around you career field of interest
- Do not use "cute" nicknames or derivatives of your military rank in your title

You have spent years in service and during that time you were often inconspicuous or simply not present in the world of social media. Not anymore; now is the time to stand out, stand up, and be noticed.

Keep the purpose of the website or platform in mind. LinkedIn is a professional site whereby people share business articles, career advice, and post job opportunities. It is not Facebook and should not be treated the same.

You may experience a sense of freedom when it comes to sharing your thoughts on social media. It is likely that during your military time you did not participate in social media or were very guarded in what you did post. Do not let this newfound freedom become a liability and in turn damage the brand you are constructing.

Social media is not your friend. In fact, if it shows a negative past, you may want to consider purging your social media footprint. This is not the time to discuss the merits of Facebook, Instagram, Pinterest, Twitter, etc. Just grasp that it is better to be smart, safe, and highly sought by others in your transition than to be seen as damaged goods.

Mastering LinkedIn

As the largest social media platform for professionals, LinkedIn is how you will present your professional face to the world. If you are not familiar with or are intimidated by LinkedIn – get over it, this platform is your new best friend!

This one platform has become a massive game changer for my business (with millions of dollars of business traced back to it) and the business world as a whole.

LINKEDIN **MARKETPLACE**

⇨ LinkedIn has 900M+ users base & it's rapidly adding 3 users every second.

⇨ LinkedIn is THE #1 JOB BOARD, hence attracting millions of job seekers and service providers alike.

⇨ Microsoft & LinkedIn joined forces because they know 3% of any given market will buy at any given time.

⇨ Buyers on LinkedIn Services Marketplace has a pain, a budget, and ready to make a decision.

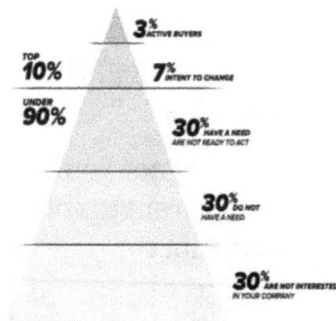

3% ACTIVE BUYERS

TOP **10%**   **7%** INTENT TO CHANGE

UNDER **90%**   **30%** HAVE A NEED ARE NOT READY TO ACT

30% DO NOT HAVE A NEED

30% ARE NOT INTERESTED IN YOUR COMPANY

There is no real one right way to build a profile to present yourself to the world. I liken it to a virtual Main Street business window, where people may look in on you without you knowing. This means you need to put your best image forward. Before you do anything, check out our LinkedIn profiles (Kevin Preston and Dr. Jeffrey Magee) as a reference. Once you've done that:

1. SIGN-UP & GET READY - Sign-up for a free profile account. You do not need to pay a large monthly fee, however, I would highly encourage you to have the minimum paid account level so you can access greater detail than you would otherwise have on the free account.

2. Before you start letting anyone know you have an account, make sure you get your first profile upload RIGHT. The beauty of your LinkedIn profile, it will always be a work-in-progress as opposed to a static page.

Understand that this is the world's greatest resume builder. It is simple. The data you put in drives how people will find you and will drive all your future searches, so be meticulous.

First, consider the first impression.

- Have a great professional face shot for the top of the page circular area.

- Use the horizontal space behind your head shot for a professional appropriate background wallpaper shot.

- Think about the ABOUT YOU space and what you want people to see you as – what are your values, mission, vision, or mantra.

Second, start building. As you start at the top in your admin dashboard, answer every field you can and add narrative under every:

- EXPERIENCE/JOB - Make sure the military labels, names, titles, and areas of responsibility are spelled out and explained in a manner that a civilian with no military experience will realize what a value add you would be to their team given what you have done, accomplished, and been responsible for.

  Search out a dozen of your peers to see the diversity and depth of how active duty professionals may be presenting themselves. Then do the same drill for anyone you know that has transitioned from the military in the past 12 months. Make sure you add the tasks, duties, and responsibilities (TDRs) of each EXPERIENCE/JOB you have or have held and don't assume the reader knows what anything means. Detail out the span-of-control you have held and the knowledge, skills, and abilities (KSAs) gained or showcased in any EXPERIENCE tenure. This is what communicates your value to others! Don't embellish and don't be humble.

- EDUCATION is critical. The military is the number one consumer of educational development on the planet, so make sure you list all of the education, license, and certifications you have acquired throughout your military tenure. While this may seem like either over-kill or insignificant to you, it is of massive value to the employer today. The market is flooded with mediocre individuals looking for a job. Remember employers may want a veteran, but they know to be selective in the hiring process to find the true rock stars that are available!

- LICENSES & CERTIFICATIONS - Anything and everything you have goes here. Seriously, you are in the armed services; there is no greater consumer and provider of schoolings, certifications, and licenses than the military. List everything here, so you can remind yourself of what you have accomplished and the value you can bring to others. Also list this so potential relationships will know the who, what, when, where, and how of connecting with you that would be of value. The more you have here provides you with opportunities to apply the VALOR System in now serving the private sector for monetization!

Next, let's get real. Fill in every section from here downward. More is better. You can always adjust or delete later. As you enter previous military employers or active duty service, your business icons will populate your page. Do the same with education, certifications, licenses, and volunteers.

With each area you add, you have the option to include clips within that specific section of work product, papers, pictures, etc. Take advantage of every opportunity to showcase your abilities.

- RECOMMENDATIONS. Remember that saying about "guilt by association?" I prefer to say "success by association." Be strategic with RECOMMENDATIONS and request them of individuals whose name, organization, or credentials/titles have power and would impress someone you want to have consider you. When you request a recommendation, it's okay to provide some helpful hints of what you'd like them to say.

Now that you have the profile set up, have several high profile and impact professionals you know vet it for feedback.

FOLLOW - Now, you are ready to start passively letting people know you are open for business. Follow others before reaching out to people to talk or invite to connect with you. This is the power of LinkedIn and aiding you in getting real time research on industries, geographies, businesses, and individuals that you may be interested in for your transition campaign.

Most every major organization, business, association, and even government agencies and each military service have official professional profiles. Go to everyone and hit the FOLLOW option on all. Now every time they post, you will get a notice and you can read up on them at your leisure. Those posts will give you valuable intel into what they are doing and even VIP points of contact. Now, you can click on any of these official posts and the person's name, and it will link you to the personal profile. You can get insights as to who they are, their background, and possible commonalities you may have with them. And, by FOLLOWING them, you will continue to see from their posts what they are doing, what they value, where you may have opportunities to connect with them, and what connections from your profile you may have back to them.

Imagine you have a presence on this playing field as far out from retirement as possible. You have massive advance opportunities to meet, network, interact, and forge relationships that can mutually serve.

CONNECT – You have plenty of time to become somewhat more comfortable with this platform. Refrain from making posts or even commenting on anyone else's posts that will draw you into a nonprofessional representation of you. Leave opinions, politics, and current events fueled by rhetoric to others!

As you find an organization of interest, go to the official business profile page, then notice the ABOUT option to learn more from their perspective about who they are. Next to that is the PEOPLE option. Click there and everyone with a LinkedIn profile in that organization will come up; here is where you can then review key stakeholders and influencers page. This will provide you with valuable insights as to how to reach out to multiple people within any organization.

Now when you want to officially reach out to them for employment investigation, you may have new advocates and allies that may provide you with valuable key insights to accelerate your successful interactions.

Connect with a personal note, not just a naked CONNECT request message.

Check back into your dashboard on a regular basis to see who has reviewed your profile. The navigational update bar is a gold mine as a people and business intelligence collector. As you see people that could aid in your transition or be actual lead generators to job opportunities, contact them to CONNECT and start a conversation dialogue.

Content is king and on LinkedIn always maintain that this is the only professional platform for business professionals doing business. This is not Facebook. Do not allow yourself to get sucked into posts, posting, or emotional rhetoric and comments. Always remember, no post is better than a bad post.

You need to be FOLLOWING substantial people and when they post, you can comment on that or share onto your own profile for your own followers to see. Look for opportunities to post unique articles or videos yourself that increase your exposure and raise your brand. This will further aid in your transition goals. Look to post or re-post daily, especially if you are reading this book and you are in the midst of your military transition period.

Now identify on LinkedIn people that you emulate, look up to, people that are where you want to be, and become a student of how they present themselves via LinkedIn. Follow them there and anywhere you can to see how they operate in the civilian world you are transitioning into and want to excel within.

Look at their LinkedIn profile and yours for areas of commonality: where you have lived, worked, went to school, if they were in the military, any similarities, etc., Then reach out to them and establish a communication and maybe even a mentored relationship and ask for their insights and guidance.

While there are a lot of resources and social media platforms you can leverage in your military career and professional career, LinkedIn is (at the time of this book) the leading tool you can use, so use it!

Position Yourself as an Expert

You may want to explore how you can leverage your expertise at your present state and become an interview guest on podcasts and/or be interviewed for publications read by those in an industry you want to enter, thus seeing you as the new hot brand and expert. Every industry has a trade association as discussed earlier in this book, and every one of them need content, speakers, and experts for their needs (meetings, communication vehicles, etc.), and these too could be targets for you to focus in on and establish a relationship with.

Remember, you are now the brand and you need to apply the **VALOR**™ model to guide you through strategies and tactics to raise your **VISIBILITY**, illustrate ways for you to gain **AWARENESS** into the marketplace, show you how to **LEVERAGE** yourself for your next **OPPORTUNITY**, while being strategic in the **RELATIONSHIPS** you have now and in your later career.

If there are YouTube posts or podcast appearances by the people you will be meeting or interviewing with, study them for context, conversation ideas, how they present themselves, and what values they may be projecting. If the business has a business website review it and their press room for additional insights to how they dress, project, and talk for gained insights to you brand alignment when meeting with them.

Now that your brand is established, it's time to focus on your resume. In the next chapter, we'll discuss the importance of storytelling skills in this step.

Call-to-Action:

Now reflect on what you have just read, ideas that have been activated in your head, and weigh them against each element of the **VALOR™ Model** and what you need to be doing right now at this moment in your career/professional life and with what makes sense for you ...

VALOR: The Ultimate Handbook on Military Transition

What are you doing and need to be doing with respect to the strategies and tactics to raise your **VISIBILITY**,

_____

_____

_____

Illustrate ways for you to gain **AWARENESS** into the market-place,
How can you **LEVERAGE** yourself for your next **OPPOR-TUNITY**,

_____

_____

_____

While being strategic in the **RELATIONSHIPS** you have now, who are they,

_____

_____

_____

_____

And who can you identify that you don't know and need to get
connected to for potential advocacy in your later career,

_____

_____

_____ !

Chapter 8

Write Your Resume

The Purpose of a Resume

A resume is not an officer or NCO evaluation, nor is it your military separation document (DD214). It is not your awards or copied statements from your evaluations. A resume, likely the first you have produced, will summarize your background, experience (tasks, duties, responsibilities/TDRs), and education (knowledge, skills, abilities/KSAs) as it applies to the specific job you are interested in. Let's be clear, a resume is a marketing document, and you are the product. Creating the best possible resume will give you a leg up on the competition.

Crafting a Powerful Resume

Following the advice in this chapter to write a resume that will be well-received by any talent acquisition officer.

Embrace Your Military Service

Veterans are sometimes advised *not* to acknowledge their military service. I am not even sure how you would structure a resume with a gap of several years to decades or using bland statements with no organization to which to attribute them. That aside, this is lousy advice!

Your base of experience is your service, and your background is very unique. Because of your service, you can effortlessly discuss:

1. Years of education and practice as a leader
2. Adept at working in extremely stressful environments
3. Global and international experience
4. Negotiation and consulting
5. Values-based leadership and decision making
6. Financial, systems, and equipment manager that quickly rises to hundreds of millions

Layer on top of that your skills and education that directly apply to the tasks of a position such as: corporate communications (public affairs), personnel operations (human resources), acquisitions (strategic sourcing), electronic technician (maintainer), and this list goes on and on.

Your challenge is that you must bring forth this very unique set of skills in a language and format that makes total sense to one who has never touched military service.

Use Civilian-Friendly Verbiage

Some of the simplest advice I received is "you must have things on your resume that we recognize." What is recognizable? For a corporate job, it can be college education or certifications. If you are in the world of human resources, The Society of Human Resources, a leading association that has two certifications in project management. Any of these items are instantly recognizable by the talent acquisition and hiring managers.

What is not recognizable? Your array of military schools, especially if you delineate them by military title. In my era of the army, I attended Combined Armed Services Staff College or CAS-

3 or as spoken CAS Cube. I doubt that anyone outside of one who served in the Army in the era would have any idea the purpose or objectives of this school. What if instead I stated: "A rigorous 6-week course focused on written and oral presentations where graduates acquire the skills to present complex topics to senior executives and confidently speak in public?"

This statement is much more relatable to one who has not served and impactful as I have found little education in the corporate world that prepares one to furnish a brief.

Early on, we talked about using appropriate language and jargon during conversations with civilians. If you stick to military verbiage, you may as well be speaking a foreign language. Your task in preparing to find a job is to select experiences from your military career and become proficient in telling your story to someone who has never served. I recommend crafting military experiences around several major categories:

1. Leadership
2. Ethics
3. Global or intercultural
4. Systems and technology
5. Managing stress
6. Decision making
7. Directly relatable experiences that apply to your job field: maintainer, leader, accountant....

Here is an example of how a veteran incorporated his experience into an interview. The question was around a GPS type tracking system. Specifically, "we use the XXXXX GPS system to track all of our vehicle movement, are you familiar with this and how it operates?"

The veteran was a weapons officer in the Navy and based on a carrier. He responded with, "I am not familiar with this exact system but in my position in the Navy I was a weapons systems officer. I was directly involved in all air operations regarding weapons and radar systems of a multi crew military aircraft."

Needless to say, the veteran more than answered the question and did receive a job offer. The veteran provided a concrete example of how he has the competence to learn their GPS system. He did not clutter the answer with branch specific jargon and in a humble fashion demonstrated he far exceeds the requirement for this specific piece of technology.

One more example of using military language and how it is often misunderstood. While attending a meeting via zoom, I was speaking about operational planning, and I used the terms **"Centralized Planning & Decentralized Execution."**

Several teammates looked at me and stated, "I have no idea what you just said." I thought I may have mumbled or spoken too fast. It turns out that this phrase, while common in the military, is not common in corporations. We had a good laugh, and I rephrased my statement. Whether you are writing your resume, applying or interviewing for a job, or actually working at the job, be mindful to use civilian-friendly language.

Use Industry-Specific Language

If you are pursuing any employment in the business sector, it will be very advantageous for you if you are conversational with the language of the industry. Just as you have a language within the armed forces followed by your branch and job skill, the industry you are seeking employment in has the same.

As a macro-category, there is a language for the business community, a language for the material handling industry, and one for being a public school teacher. Regardless of your industry, being familiar with that language and providing examples of your armed forces experiences using their language is a powerful tool.

The next step in language learning is to learn the language of a specific company or organization. When you can speak with an industry language and apply the language or jargon of a business or organization, your story and brand become quite compelling. You may ask, "how does one learn the language of a specific company?"

First, publicly traded companies have a host of information that is readily available:

1. Attend or listen to their quarterly earning calls. You will hear from the CEO and CFO on priorities, success, and new initiatives.

2. Follow the company's stock via their ticker symbol; you will see a variety of articles associated with the company.

3. Read the company's Form 10-K, a document that the SEC requires all public companies to file each year. This form presents the financial picture of the company, detailing its revenues, assets, and liabilities for the previous year.

4. Take advantage of published books. Many older companies have many books that discuss the founder and their journey towards success. McDonald's for example has many books. "Behind the Arches" is an example. The movie "The Founder" with Michael Keaton presents the story of Ray Kroc.

5. Read job postings, regardless of whether the position applies to you or not. Job postings will often present a history of the company, define their purpose and mission, and they will use their company language.

6. Become a regular student of Netflix's MasterClass series and listen, learn, absorb, and then use some of these powerful business leadership life lessons shared.

These sources can prepare you for the submission of a thorough industry and business specific resume that will demonstrate your knowledge and preparation for the position of interest.

Communicate Your Accomplishments

In the military, your uniform is your resume. By that alone, one can tell your level (rank), schools attended, accomplishments, tenure in the organization, and special recognition (ribbon rack). However, you will not wear your uniform while seeking employment outside of the military (and a civilian employer wouldn't understand it anyway), so you need language to convey your accomplishments and awards.

You do not want to list all of your awards on your resume. It will make absolutely no sense to anyone outside of the military. Instead, convert those awards into relatable language and place them in their own section on your resume.

Topic heading: Awards and Recognitions

Statement: Recognized xx times by The United States Army (insert the number of awards you have, include devices) for exceptional leadership with the highest award being the: (Bronze Star, Meritorious Service Medal or DCS...)

This method is clear and relatable to anyone in the hiring process, and it's not cluttered with award abbreviations. If by chance you are speaking to a person who has served, there is still an opening for conversation.

By doing your market research you can explore and identify market opportunities. By then reflecting upon your network and more importantly the relationships that you have, you will be able to determine which doors will be easy to access and which doors may be more difficult to get opened for you in your journey.

A simple chronology may look like ...

1. Your resume serves as the document to highlight your best qualities and experiences, which will capture the attention of the recipient. This will ideally drive them to want to meet you ...

2. With your resume in hand, they may be compelled to re-search you on the internet. This is why we have spent so much time discussing social media, LinkedIn, and your online footprint ...

3. The narration of your professional life history will be critical in getting market share in someone else's head and thereby making them want to meet you ...

You don't have to put everything in your resume. Save some elements for the first interview discussion. We'll talk more about the interview in the next Chapter.

If you have a comprehensive LinkedIn profile, and you have expanded the text under every experience/job you have held and have completed every section within the administrative dashboard in building out your profile, there is an option to have a resume built from your profile information. This is very impressive as an option as well.

Save the Humility for Later

You have built your career on a platform of teamwork and co-operation. Over my time of service, I found individual recognition to be a bit uncomfortable. Despite awards being for the acts and accomplishments of the recipient, I always felt that a team accomplished a task, and being acknowledged for the team always felt a bit awkward. Humility is a wonderful trait and has and will serve you in the future. However, during the job search process it has no place.

Always keep in mind when an employer is filling an open position, they want to know what **YOU** did and how **YOUR** experience applies to this position. Here are a few quick tips:

- **Use "I" instead of "we:"** For example, I designed and implemented an inventory management system that eliminated duplicative purchases and saved $10.2 million annually in inventory management cost.

- **Practice, practice, practice:** While serving, you became very accustomed to giving credit to others. Though this is a very positive trait, you need to become conversationally comfortable in presenting your accomplishments. Develop a script of many topics that can be applied to your resume and presented as examples of your perfect fit. These points become your talking points when you reach an interview.

- **Acknowledge your focus on a team:** In a world full of people who want to take credit for everything, be the one who is focused on developing the team. One of my favorite quotes from President Reagan is "Imagine what we can accomplish if no one is concerned about who gets the credit." This may sound like a departure from what we just discussed; when presented properly, you can be the person who coaches, mentors, and develops those around you; simply stated, regardless of title, you are a leader.

Allow Plenty of Time to Test Drive

Make time your ally and not your enemy. When this philosophy is applied to a resume, assure you have time to have this reviewed and corrected.

I recommend crafting a solid general resume, one that is not written for a specific career field. This resume is clear of military jargon and language, and if it is used, you define it. You strategically use the "real estate" of the paper, and it is prepared in a pleasing, simple to read format. Once you have built this quality document, you can pivot to job specific resumes.

One of my favorite methods to test drive your resume is to find a friend who has never touched the armed services. Present your resume and give them a highlighter. Ask them to highlight anything that does not make sense to them. Take this into account when creating your 2.0 version.

You can also call on colleagues from the business community. I would do this after you have had several reviews and your resume is generally de-cluttered from military language. This will serve two purposes, one you will have another set of eyes that can provide industry specific advice, second, this is an avenue to socialize you and your brand within a company.

I have had many times where someone from talent acquisition contacts me asking if I know anyone for a role they are trying to fill. I will reference resumes that I have on hand and send any that fit the request. This is a tremendous avenue to have your resume submitted with a possible endorsement from a trusted company. This is not something I would request; it's better to let this emerge organically as the relationship matures.

You can create and have your resume and supporting document packet edited by your board of directors and advocates, then use your standard resume as a template. For the targeted business that you will submit your credentials to, you can then study that organization's language, culture, values, and goals and then tailor some of your content so that it speaks to them and their language. Present yourself in such a manner that the reader sees you as a member of their organization already providing value. This will drive your name to the top of the candidate consideration list.

Call-to-Action:

Now reflect on what you have just read, ideas that have been activated in your head, and weigh them against each element of the **VALOR™ Model** and what you need to be doing right now at this moment in your career/professional life and with what makes sense for you ...

VALOR: The Ultimate Handbook on Military Transition

What are you doing and need to be doing with respect to the strategies and tactics to raise your **VISIBILITY**,

_____

_____

_____

Illustrate ways for you to gain **AWARENESS** into the market-place,
How can you **LEVERAGE** yourself for your next **OPPOR-TUNITY**,

_____

_____

_____

While being strategic in the **RELATIONSHIPS** you have now, who are they,

Kevin Preston & Dr. Jeffrey Magee

_____

_____

_____

And who can you identify that you don't know and need to get
connected to for potential advocacy in your later career,

_____

_____

_____!

Chapter 9

Finding Opportunities

<u>Now, the real selling begins!</u>

You do not need to wait until your military career separation is on the horizon to begin looking for opportunities. Knowing what's out there will help you make an educated decision and find the best employment possible. Look for targeted on-line portals and businesses where you can submit your credentials and applications for employment. You may even want to test the waters by applying for a few positions.

<u>Narrowing Down Your Focus</u>

There are endless opportunities out there, which means you will have two important tasks during your job search.

First, divide these many career activities into two categories to narrow in on your career field.

1. The roles and responsibilities that you truly enjoyed, and that you have demonstrated competency coupled with experience. In other words, you have a passion for this work! When leaving military service, you have accumulated a very broad array of skills. You have specified skills for which you attended a lengthy school, for example an intelligence analyst is a four-month school, and you now have a specific

identifier. You will have additional duties, some of which may have a school associated with and you have career changes of duty reassignments that may produce yet another skill. You must be able to demonstrate your expertise through years of associated projects.

2. Those roles and responsibilities you did not enjoy or do not have the proper experience to carry out. For example, having served as the physical fitness NCOIC/OIC 15 years ago is likely not a level of experience that will qualify you to pursue a position as manager of corporate wellness programs.

Once you have a specific focus on a career field, your next step is to pursue this with unending tenacity. If you served as an intelligence analyst in the military, you have 10+ years of experience, have completed many schools, and you truly enjoy this work. You will apply all of your efforts in this direction. What you **do not want to do** is apply for corporate intelligence positions, marketing positions, public affairs positions, and human resource positions. When you have this scattered approach to your future, it dilutes your application and brand.

The talent acquisition team will see the many applications across 5 very different business functions and will quickly draw the conclusion that you have no passion about any one field. When hiring, people seek experienced applicants with transferable skills who can integrate with a team and have passion about the career field. A scattered approach and applying for 5 positions across 5 career fields does not support this strategy.

Second, you must get clear on the level of work or the title of the position you'd like to get. (The military equivalent is your rank or rating). Earlier, we mentioned that you should not apply for jobs across different career fields. Now, we want to caution you against applying for positions with vastly different levels. This would look like putting in applications for a vice president, one for a manager, and one for a coordinator. The years of experience between these roles can be as broad as 20 years. It's the equivalent of experience between a specialist (E-4) and a master sergeant (E-8).

Quite simply, you want to be extremely focused when selecting the career field and the level for which you apply. When exercising this control, you will present as focused, experienced, and passionate, all items that employers are seeking.

Research Potential Industries

It is important that you do your homework here. Learn about the industry or potential company you want to work for or within. Research the industries and/or business of interest to you. You can even find potential employers within an industry that you may not even know exist.

The federal government has assigned a number matrix to every industry and within every industry, every business. Finding out the codes for the industries you are interested in empowers you to find potential employers anywhere in the nation.

- Standard Industrial Classification (SIC) codes are four-digit numerical codes that categorize the industries that companies belong to based on their business activities.
- Standard Industrial Classification codes were mostly replaced by the six-digit North American Industry Classification System (NAICS).

Armed with this data, you can search by industry, by trade, and even by geography for employment opportunities that others will most assuredly miss.

You can learn more about transition career opportunities through their respective trade associations as well. If you are still in the military, you should have long since been involved in these. If not, you are behind the curve and should consider involvement endeavors right now. For example, if you are in the Human Resources area (G1), then your civilian professional trade association would be the Society for Human Resource Management (SHRM), SHRM - The Voice of All Things Work. Within that organization are national involvements, meetings, causes and certifications. Every major city in America has an operating SHRM Chapter, and here is where you can meet and network with your affinity group, tribe, contacts, etc.

There is a professional trade association for every position within any organizational structure. Seek out those that are relevant for who you are and where you are now, then do the same for where you want to be. Whether your interest is in the trades, blue collar, white collar, or whatever designation you wish to use, there is a group out there representing it globally and you need to be aware at the least, and preferably involved.

Here are some of the associations you may want to look into:

1. Accounting - aicpa-cima.com
2. Legal - AmericanBar.org
3. Talent Development - td.org
4. Subject-Matter-Expert speaking for a living - NSASpeaker.org
5. Engineering – nspe.org
6. Project Management – pmi.org
7. Auto mechanics – amra.org
8. Supply Chain – ascm.org

9. Medical supplies – supplychainassociation.org

10. Nurse, Radiologist, Doctor – AMA.org

11. Every MOS, AFSC

Stay abreast of your targeted industry(s) and potential actual employers online by following them through other online employment hiring portals such as: Indeed, monster.com, Robert Half, Express Personnel, etc. Look to see how they use those portals to find candidates like you.

Research Potential Companies

Intelligence Preparation of the Battlefield (IPB) is a term familiar to those who served in the Army. IPB integrates information on the enemy, weather, terrain and provides a basis for situation development and the formation of courses of action. In short, a commander wants to know what is on the battlefield and what is the enemy's likely course of action.

When you do IPB for job applications and interviews, you will look to the following sources of information:

- LinkedIn where you can follow brands, businesses, associations, key personalities
- Annual report, 10-K, shareholders statement
- Their social media footprint
- Trade journals, professional pubs.
- Podcasts
- Books their employees have authored
- Employment postings

In my leadership book *THE MANAGERIAL LEADERSHIP BIBLE, Third Edition* and graduates of my *LEADERSHIP MASTERY:*

*LEADERSHIP ACADEMY OF EXCELLENCE/*1.0 &2.0 Series, I discuss how everyone who is successful in business has a specific critical ability...

Situational Awareness =

Strategic Effectiveness™

Your job is to apply this philosophy to your future employment.

A company's position postings are a map into the language and culture of that company. Most position postings will provide a brief company history, describe the mission, and provide a glimpse into the culture. You will also learn the language used by the company. In my case, words like magic, guest, dream, and cast are commonly used at Walt Disney World. Regardless of whether you blend these into conversation or not, being comfortable in company and industry language will boost your confidence.

My (Kevin's) path for preparation to transition was an extended period of learning about my desired company. Over several years, I engaged in the following:

1. Read numerous biographies and books about the history, the founders, and mission of the company
2. Listen to quarterly investment calls
3. Reviewed the Form 1-K - a comprehensive report filed annually by a publicly traded company about its financial performance
4. Learned the operational layout of the company
5. Read employment postings
6. Consume the product produced

By the time I applied for a position, I had a thorough and conversational knowledge of the history, operations, finances, mission, and purpose.

Understand the Application Process

Most organizations have an official front door for applying (human resources department or a recruiter within that business unit). Always respect and work through them if that is what is expected. However, there are other ways to be noticed when job hunting. Ask yourself:

- Where can I best bring value to this organization, what business area, unit, department?

- Who would be the senior most manager/leader in that area?

- Who would actually be the supervising manager?

- How do I identify that name and research them via their website or LinkedIn profile to explore places of connectivity or connection?

- How can I best get connected with them, initiate a discussion, and explore getting my name in their head as a new addition to their team ...?

Sometimes identifying the supervising manager over the hiring manager can get you hired when a hiring manager or recruiter may have not even known that there was a need that they should have been searching for.

Identify, assess, and connect with ALL military-veteran-employment groups (such as HiringOurHeroes.com), associations, service-based groups, chambers of commerce (and recognize that there are chambers of commerce by both geographical area [national, state, regional, metro] and by ethnicity [Latin, urban, Jewish, etc.] as well as by industry, etc.), where you can learn of hidden opportunities and introduce your brand to those who do not know you.

Explore organizations that may have an opportunity for you to interact through a volunteer entry area or as an untraditional Intern or part-time employment in a seasonal situation. There are endless creative ways to test the water.

Leverage online and brick and mortar employment search firms as a part of your new front-line soldiers to represent you. Your ability to know where the opportunities will be limited based on your own awareness and efforts, now imagine the outreach that can take place when you can tap into deeper and wider networks looking for great talent – just like you!

Making sure that your credentials are logged in with multiple other sources, where employers may cast their net wide and far, you want to look at the market as a global market.

1. Veteran owned and operated employment search firms and recruiters
2. Non-veteran owned and operated employment search firms and recruiters

And, then from here recognize that this business also may have greater effectiveness for you by:

1. Government contractor niche and contacts
2. Non-profit niche and contacts
3. Private business sector niche and contacts
4. Public business sector niche and contacts
5. International niche and contacts
6. Domestic niche and contacts
7. Regional niche and contacts
8. Industry niche and contacts

9. Skill/competency niche and contacts

10. Labor intense niche and contacts

11. Etc. ...

So, when it comes to getting your name out there and applying for opportunities, there are more opportunities than there are viable candidates, so stay massively optimistic and recognize that now it is about positioning and selling YOU as the brand.

<u>Apply</u>

There is one school of thought where you send your resume to any company and any position looking for a warm body. This shotgun approach wastes your time, the recruiter's time, and can actually backfire as you won't be seen as a serious candidate. Be honest and transparent. Recruiters will see all the positions you applied within their company, so if there are multiple opportunities within an organization of interest to you and which your credentials support, use this as a reason for a pre interview conversation with the recruiter, hiring contact, or organization.

Leverage your connections to determine which job may be best suited for you in the present and which may lead to the best opportunities in the future. You may also have a relationship with someone on the inside who can advocate for you with hiring managers.

If you do apply for multiple positions, make all parties aware as you move through the interview process. If you are withdrawn from consideration for a position, be proactive and let your recruiter know. This will allow you to control the narrative and avoid awkward conversations.

Now the real selling begins. Here is where you launch your VALOR Campaign. The follow-up touchpoints must maintain and raise your **VISIBILITY**, illustrate ways for you to gain **AWARENESS** into their business with all decisions makers and influencers to hiring you and promoting you once within the organization, **LEVERAGE** yourself within your network, community, and connections for this **OPPORTUNITY**, while being strategic in the **RELATIONSHIPS** you have now to help you.

Understand Compensation

Let us be clear, your job search should not center on the money you will make, and you should **never** discuss compensation until an offer has been made. The reason we are sharing this information at this point in the book is because it's important for you to understand how civilian pay differs from military compensation.

I made this mistake early on in my planning to retire. I was very focused on compensation. I would classify myself as no more or less motivated by money than the next person, but it became the foundation of my search. As I reflected on "why" my search for compensation was driving my behavior, it dawned on me. The only item I understood that was common between military service and the private sector was salary. Given this equivalency between the two, that is what I structured my interactions around. This was not a good method. Employers are looking for a person who has a strong passion for the field and a genuine enthusiasm for the company, not someone who is "just seeking a paycheck."

Thankfully, I received some liberating advice from a retired sergeant major. He said, "The money will find you. You do not need to chase it." We discussed how a purely compensation motivation comes across to a hiring manager, and I understood the error of my ways. The sergeant major's advice was exactly correct and freed me to pursue employment that excited me with companies that I respected.

Calculate Your Worth

It will also be helpful to calculate your worth in the civilian marketspace. What does your entire military compensation look like? This could include salary, health care for you and your family, work hours, special prices on commerce (PX, Commissary, Gas, Recreation, etc.) This will give you a better holistic idea of what you would need as a minimum compensation in your transition.

Consider whether you are transitioning from the military with a retirement and health benefit package. If you are exiting the military without a retirement or health benefit package, you really need to stop and consider these factors. Since President Obama's Affordable Care Act went into effect, the cost of healthcare has exploded to rates no one would have even conceived – this has been a disaster on the civilian economy. For example, if you are 55 years of age and in good health, to self-insure with a basic health care policy would be upwards of $800.00 a month (as of 1-1-2023).

Not all employers offer healthcare benefits or offer limited plans. If you do not have your own, you will need to factor this possibility in when choosing a company to work for. If you are transitioning without retirement, then also explore with the civilian opportunity or government contractor job if that is your route, what tenure they may give for your time in service or not. What does their retirement incentive program involve and how does that play for you and your transition years?

Kevin Preston & Dr. Jeffrey Magee

Health Care Benefits

Over my many years of service, not once did I ever think about health care. One of the tremendous benefits of serving is quality healthcare coverage for the service member and their family. Simply stated, healthcare was always there and was part of my service agreement.

When I retired, I entered the TRICARE retired program. Let me describe this. My wife and I have a local primary care physician, he monitors our health and recommends various referrals as needed. To this point, I have not used the Veterans Administration for health care. As a retiree from the Army, annually, my wife and I pay less than one thousand dollars for our health care and the quality we receive is top notch. This, compared to any private coverage, is very inexpensive. Simply stated, you have an incredible benefit when your military health care coverage follows you. One last note, when you keep your TRICARE coverage, this provides you with a tremendous amount of flexibility. Your health care is not tied to your employment, you can pause work, retire early (before Medicare eligibility), open a business all with the assurance that you have quality health care coverage.

Factoring In Your Military Retirement

A common question for anyone leaving service, regardless of if they are retiring, receiving VA disability compensation, or going to serve in the Reserves or Guard is how will my employer treat my military income when considering their compensation package?

At one point in my preparation to transition, I viewed my military retirement as a portion of my compensation. Let me explain. I used my projected military retirement as a percentage of my income and my approach was, I need to cover the difference with my next employer. Therefore, I would be "whole" in terms of my income. WRONG, WRONG, WRONG!!

Any compensation you receive from your military service is independent from your future income.

Your next employer is paying you for your experience and for the work you will perform. It does not enter into the calculation of your compensation; in fact, *view any military annuities as payment for the work you have already performed.*

Calculus of Salary

It bears repeating: Do not discuss salary during the interview. However, if the employer brings it up, they may ask, "How much are you seeking in compensation?" Unless you are really certain of the parameters and range of compensation, I would not immediately answer this. Instead, I would reply, "I would like to understand the full scope and responsibility of the position then I can better furnish an answer."

What you do not want to do is give an off the cuff answer and in turn significantly short yourself on compensation, if this is your first job following service, odds are you will short yourself as you have no idea what the market pays.

While military compensation is incredibly transparent, the private sector is generally not as forthcoming. While in the military, everyone knows or can calculate anyone's salary, but no one cares. In the private sector, compensation is generally a very guarded conversation and not an item discussed publicly.

Knowing this, how does one navigate salary in the job search process? While the private sector does not have the published pay charts as does the military, there are still avenues to conduct some research. The trend of late is more and more private companies are posting a salary within the job announcement. This can be a big help as you will know up front if this is within your acceptable range. You can also speak with your board of directors and search the internet. There are online sites like www.Salary.com where you can identify a similar job and responsibilities to what you are doing within the military, add the zip code where you would ideally like to work, and it will share with you the range and mean pay scale opportunities. Armed with this information, you can provide a well thought out response and one that is appropriate to your experience and background.

Negotiating Your Salary

If salary never comes up during the interview, the offered salary will be made when they offer you the job. It is perfectly acceptable to request 24 hours to discuss this with your family. Take this time to consider all the elements of compensation: salary, bonus (if it applies), paid time off, benefits, company perks, and any other parts of the offer.

First, make sure you respond well within the 24-hour period. Failing to respond within the allotted time can send a poor signal and may result in them offering the position to another candidate. Depending on your level you may have some latitude to negotiate a higher salary but also keep in mind, they likely have a second and third choice identified.

You can benchmark your salary and compensation offer by using online sites like www.Salary.com or reflecting on the TDRs and KSAs that you possess and thus the VALUE you bring to a perspective new employer. Tour value is what you are monetizing, look at others as best you can that are working within your targeted organization, and reflecting upon what they bring to the organization by way of talent and what you are bringing.

You have traveled a tremendous distance in landing your perfect job with your desired company in an industry of interest. This is your first foray outside of military service. It would be a real shame to let a difference of a few thousand dollars derail the many months of hard work. You know your financial needs, but a place of employment goes beyond that dollar and considers culture, your enjoyment of the people on your team, and many intangibles. Whatever your decision, stay true to your word and be honest throughout the process.

Call-to-Action:

Now reflect on what you have just read, ideas that have been activated in your head, and weigh them against each element of the **VALOR™ Model** and what you need to be doing right now at this moment in your career/professional life and with what makes sense for you ...

VALOR: The Ultimate Handbook on Military Transition

What are you doing and need to be doing with respect to the strategies and tactics to raise your **VISIBILITY**,

_____

_____

_____

Illustrate ways for you to gain **AWARENESS** into the market-place,
How can you **LEVERAGE** yourself for your next **OPPORTUNITY**,

_____

_____

_____

While being strategic in the **RELATIONSHIPS** you have now, who are they,

_____

_____

_____

_____

And who can you identify that you don't know and need to get connected to for potential advocacy in your later career,

_____

_____

_____ !

Chapter 10

Ace Your Interview

If You Don't Put Your Foot in the Water, You Will Never Be Able to Swim!

As a military professional, your career has been on a set trajectory, making interviewing and competing for a promotion unnecessary. As you approach the end of your service, it is time to start exploring entry points to the civilian job market. In order to do this, you'll want to make sure that your interview skills are top-notch, and when you do get the opportunity to sit across from a potential employer, you wow them.

Preparation is key. That's why this chapter is broken up into four sections:

1. Before you apply
2. Prepare for the interview
3. Step Into the interview
4. Follow up

Before You Apply

Have an Interview Advocate – This may be one of your board of directors or someone you have a valued relationship with who knows you and may have an understanding of the business you wish to interview for. They can serve as an advocate or sort of interview coach to what you need to emphasize about your background, ways to story tell that, and what topics you should avoid. Ask them what are the organization's immediate and long-term needs in talent that you may be able to address and thus bring up in your application, inquiry letters, and follow up letters as a way of positioning yourself as the #1 candidate to be hired.

An Interview Advocate is ideally someone with direct knowledge of the organization you want to interview into and may even know some of the key stakeholders within the targeted business. They may also serve even deeper as your advocate from the inside to key stakeholders making decisions and they can present you, defend you, champion you above other candidates that may also be being considered for the job opportunities.

Schedule a Mock Interview

Many times, professional employment search firms have mock interview sessions and training for their clients before they send them out to an interview appointment that has been set. You may want to explore local search firms and see how they work and if this is a service they provide. If so, immerse yourself. If you are not working with a firm, ask someone on your board of directors to work with you.

Do you have a select industry or actual business that you would like to interview with? Even if you won't exit the military for many years, you can begin evaluating your existing relationships now. Is someone you know in a place you'd like to be? If so, ask them if you could do a mock interview with them.

Use this mock interview opportunity to see them as the hiring manager and not just as your friend and go through the experience. Afterward, ask them for an after-action review or what the civilian world would call a debrief. See what points you can glean and improve upon. Ask them:

1. How did you first present yourself for the interview (whether face to face, over a video communication platform, or telephone)?
2. Did you ask enough or meaningful questions and how did you respond to the questions posed?
3. How did you physically appear and present yourself?
4. Was there a clear call to action leaving that initial interview in regards to next steps for you and for the interviewer?

Prepare for the Interview

Now, you are probably wondering: "Is this preparation really important? Why can't I just go into an interview and simply respond to questions with my background?"

Several items come to mind for the merits of preparation.

1. When you are well prepared, you present in a more confident manner. Your attitude goes from "I hope they don't ask me XXXX questions," to "Bring it on, I am ready!"
2. When I was interviewed, I was asked: "Kevin, I have reviewed your resume, it's really impressive, I am not even sure what to ask you. But I do have a problem; how do we tie this new veteran's hiring program with the history of the company or do we just skip the history and start from the present?"

This was a question that falls into the "bring it on" category. I knew the company's history. I was conversational in the intersections with the United States Armed Forces and how that intersection made a positive impact. So I told him, "Embrace the past and build from that work to the new veterans employment program." I added a great deal of detail while summarizing close to a hundred years of company history, and I did this in a confident and conversational fashion. There in is the value of preparation.

Understand Who You Are

Spend time really evaluating your military career journey for the knowledge, skills, and abilities acquired and vetted through the tasks, duties, and responsibilities unique to you and craft the stories around how the appropriate responses would resonate in their head as the listener. Crafting stories around experiences that apply to the new position will set you apart from other individuals interviewing for the same position or even a position you create for them.

What you don't want to be or come across as is the scary veteran. You were in the people business and your next job is about the people business, so humanize yourself to them and humanize what you have been doing so they can relate and understand.

Understand Who They Are

Your entire military career has been predicated upon getting information and intelligence and then reviewing it for what makes the best sound judgements. Your interview is just another example of that opportunity.

Prepare Stories to Answer Their Questions

During the interview, you have the opportunity to bring your military career to life using storytelling.

You will be asked to illustrate your values, your problem-solving skills, and a variety of other desired characteristics. Using a story will not only help you evoke emotion in your interviewer (which will help create a connection), it will also allow you to explain how the tasks you performed during your military service can translate to the civilian world.

For example, during an interview, a veteran was asked to describe a stressful situation they faced and explain how it was handled and resolved:

> While serving in Afghanistan, we routinely had to travel between bases via the roads. We always traveled in convoys, a collection of 4-6 military vehicles, and were taught that once a convoy starts, you don't stop, and you do not leave the vehicle until you arrive at your destination. Our vehicle was a HMMWV. You have probably seen these in the news, a very large vehicle but the interior is quite small, especially when you wear a lot of gear. This vehicle had several wires that would work their way loose causing the vehicle to stall.
>
> One day, we were traveling through a very unsafe part of the country when the vehicle stalled. We were stuck and became a stationary target. I had to do something quickly; this was endangering the lives of everyone traveling with us.
>
> In a very small space, I quickly twisted my body so I could be underneath the passenger dash. Despite the size of a HMMWV, this space is really small. Using my multi-tool, I was able to reattach the wires in short order. Our vehicle restarted and the convoy proceeded to travel safely to our destination.

This is a brilliant story in many ways:

1. It links his military service to his job pursuit.
2. It's very relatable; this is equal to a car repair but done while operating it in a very dangerous environment.
3. It demonstrates quick thinking and extreme calm under pressure.
4. It does not use any language or imagery that is unrelatable or off putting.

And yes, he was offered the position.

The art of storytelling is a gift for many and a craft others study.

Sheryl Green, author of "Once Upon a Bottom Line: Harnessing the Power of Storytelling in Sales," breaks down the essential elements of story:

1. Purpose: why are you telling the story? (This may come at the end)
2. Life as it was: the setting, the characters, and enough back-story to understand the situation.
3. Boot to Butt Moment: something happens that disturbs the character's peaceful existence.
4. A Goal: what does the character want to achieve?
5. Conflict: What's standing in the way of their goal (this can be both internal and external)
6. Attempts: what does the character try in order to reach their goal?
7. Solution: how do they eventually reach their goal?
8. Life As It Is Now: what does life look like now that they've achieved their goal?

Study old videos online of great orators and see how they set up a story, deliver the key lines, and exit back to the listener. How they make it relatable for others. Some of the greats coaching clients on storytelling to follow, study and find ways to connect with today:

- Patricia Fripp – www.Fripp.com

- Darren LaCroix – www.DarrenLaCroix.com

- Kelly Swanson – www.MotivationalSpeakerKellySwanson.com

These are a few of the mentors I have had and have in crafting great story. Now become the Sundance Film Festival producer of Brand You!

Step Into the Interview

As we have stated, there is a fine line about advocating for yourself and being braggadocious, and it is always uncomfortable to advocate for yourself. When you meet with an interviewer, listen a lot, talk appropriately, be conversational, and always consider as you hear the questions and rapid fire scan your life history and immediate military career. I have learned all business comes down to four critical factors in evaluating human capital. The civilian employer is looking for one of the four core value adds in hiring YOU:

- How can you make their business and the team you would be a part of BETTER?

- How can you make their business and the team you would be a part of FASTER or more efficient?

- How can you make their business and the team you would be a part of DIFFERENT in a marketable way?

- How can you make their business and the team you would be a part of more COST EFFECTIVE?

Be careful to not fall into the trap of responding or saying, "I can do anything you like." Consider what you may be asked to do or if you can do it and consider the elements of the *Player Capability Index* presented in earlier chapters. If asked, can you do something, recognize if you really have the "T" training or "E" experience in something similar, and if so, then the answer is yes. If not, then respond with a sincere desire to learn, study, and become credentialed in this if hired. Be uniquely you in everything you present and speak.

Ask the Right Questions

In an employment interview, you must be evaluating the potential employer just as they are evaluating you. Come armed with a series of questions that will help you understand the company, the culture, the role, and the people you will be working with.

Here are a few questions to get you started:

- Does the new employer have a uniform expectation or allowance for employees?
- Is there driving associated with the job, and if so, do they provide a vehicle or travel alliance? Is that for your vehicle or theirs?
- What does their annual work calendar provide for paid days off, sick leave, vacation days, etc.?
- If you are excited about the opportunity, ask them what the next phase looks like. Make sure you clearly know what the next Call to Action step is when you leave the interview.

Interview like your life depends upon it, as it just may!

I (Kevin) can distinctly recall my first in-person job interview in 30 years. I was still on active duty, and I had not yet decided to retire. Up to that point, I had a series of phone meet and greets and interviews, and I had advanced to the last round: the in-person interview. I was very excited and very scared.

My interview was scheduled. The company paid for the airplane ticket, rental car, and hotel room. As I mentioned in the Brand Chapter, I knew I needed appropriate attire. I had a full military wardrobe, but that was inappropriate. Off to Nordstroms I went.

The day before, I set out to recon the location, driving from my hotel, determined the travel time and made sure I knew where to "report." I woke up early the following morning, did a quick equipment check: attire, challenge coins, business plan on "the merits, steps, and strategy for building a veterans program," and my aviator helmet briefcase. I set out for the day ahead. I felt prepared.

My first stop at the company was security. I checked in, presented ID, and waited for a shuttle service to drive me to the building where I would meet an array of individuals. While driving, I marveled at the beautiful campus and plush work environment. I was dropped off at building seven, entered the lobby, and I was met by a gentleman named Mike. I had spoken to Mike through the interview process so there was a familiarity. Mike was not part of the formal interviews but rather the guide that moved me from person-to-person. Though Mike was not interviewing me per se, he was influential.

Mike had prepared a schedule for the day. This was helpful and comforting; I now had a "road map" to follow. My first meeting was with a gentleman named Chris, head of corporate communications. Mike walked me to a small conference room, invited me to sit and wait for Chris. Chris entered the room, I stood and shook his hand, his first statement was "I have read your resume, it's very impressive and frankly, I do not even know what to ask you."

We stared at each other for a moment, and then he said, "With this new veterans program, how do I blend and communicate our history with the armed forces and tie that to the present?"

In my preparation, I had read many books on the history of the company and the intersections with the United States Armed Forces. I felt well equipped to answer this question.

We spoke for about 45 minutes. A few takeaways from this meeting:

1. I came with relatable stories from my military experience.
2. I spoke in sound bites, placing the headline at the top and filling in underneath.
3. I was aware of key speaking points that I wanted to get into the conversation at the top of the meeting. (You never know how long a person can stay in a meeting, always place your critical information at the top of the meeting).
4. At the close of the meeting, I gave him a challenge coin and said, "Thank you for what you have and will do for the veterans of this country." I presented every person I spoke to with a coin.

At the close of this meeting, Mike returned and walked me to the next office. The next meeting was with Jackson. He led Diversity & Inclusion. When we arrived at Jackson's office, Mike knocked on the door, opened it for me, and Jackson invited me to take a seat. This simple gesture may seem insignificant, it was not!

Left to my own devices, how do I enter a closed office? I know I do not treat it like I am in the Army: no loud knocks on the door, no loud verbal statements. Fortunately, Mike eliminated that problem, knocked on the door, opened it, and introduced me to Jackson.

My conversation with Jackson revolved around human resource and care-oriented issues. I felt well equipped, I discussed Army family programs and work around soldier resiliency. I discussed the significant role of the military family and how this well-functioning support system makes military service possible. We talked about military family care and the importance of military spouse employment. In my aviator helmet bag, I happened to have a reference guide that I built for the military family that served as a quick resource. I gave Jackson that guide. I left the interview feeling confident that all went very well.

Some takeaways from that meeting:

1. I worked extremely hard to be ingratiating and approachable. That translates to smiling, not being so stiff and rigid, not staring, and not being a typical dominating, intimidating soldier. I did not come off as the "scary soldier."

2. I was well prepared with relatable experiences from my time in the Army, and I was able to tell these stories without the use of acronyms or jargon.

3. I was very conscious about the terms of "we" versus "I." Coming out of the military, part of our culture is that of a team. Consequently, the word "we" is frequently used. Remember, a company is interviewing you and wants to hear about what you have accomplished.

Next on the schedule was a virtual interview. This was in 2012, well before our current world of zoom. My overall experience in interviews was minimal but my experience in virtual interviews was nothing. I was led into a conference with a rather large screen in front of me and the meeting was opened. I was then left waiting for the interview to begin. A woman came on screen, I believe she was a Vice President, and we started with the normal series of introduction questions.

1. Tell me a bit about yourself?
2. What experience can you offer?
3. What unique characteristics do you bring to this position?

As we talked, I felt that the energy and emotion of an in-person meeting was lacking. It was no reflection of the person I was speaking to but rather the virtual nature of the interview. The interview was coming to a close. We had 30 minutes scheduled, and I felt I needed something to bring some excitement. The interviewer said, "I have really enjoyed our time, do you have any final questions or comments?"

This was my chance to close on an energetic high note. I stood up, looked at the camera and said, "I enjoy this work! My years of service in the Army have prepared me to step into this position and rapidly make an impact."

The simple act of standing and speaking brought a new dynamic and energy. I was able to gesture and move a bit. It changed my entire presentation. I am not sure I would recommend this if this meeting was in person; however, given the virtual nature it worked very well. I left this interview session feeling confident that I had conveyed my experience and passion in the best possible manner.

My next stop was with the Senior Vice President; this was the final formal interview. The same sequence applied: his assistant opened his door and I entered. I stood behind the chair, somewhat at attention. His first statement was "welcome Kevin; just relax, we are just going to have a conversation." That was a tremendous statement that somewhat put me at ease.

I continued with my past approach:

1. Demonstrate a passion for the brand and company

2. Illustrate how my military experience directly relates to this position

3. Highlight my relationships within DOD that would be beneficial

One of the discussion items of the interview was a strategy for the new company-wide veteran's program. I recall stating that we need a strategy that is focused and finite, one that accelerates the objectives of the company and serves as a baseline from which we can compare alternatives for allocation of time and money. That statement came directly from my recently completed MBA, yet another intersection with my preparation to transition.

I presented my business plan; I would caution this as an action. If you are comfortable with the company and competent in preparing this document, it can be helpful. It is a tool to demonstrate your knowledge and acumen. But if you miss the mark, it can also substantially decrease the chances of being selected.

Before leaving the Senior Vice President, I said, "Regardless of your decision for my selection, I appreciate the work that you will do on behalf of America's veterans. Thank you. I presented him with my coin in our final handshake and explained the nature and significance of the coin.

With all of the formal interviews complete, Mike offered lunch in the company eatery. Now despite all of the formal work being over, I realized that I was still being evaluated. Lunch in a company cafeteria is a tremendous opportunity to meet a variety of people. One person we met led their veteran-focused employee group. We chatted for a bit, talked about life in the military and I gave him a coin.

The day ended with a ride from security back to the check-in building and off to the airport. Reflecting on this day, this was the culmination of almost 5 years of preparation, and I left feeling I had put my best foot forward.

To summarize my experiences and my advice:

1. **Be Prepared:** Make sure you are conversational with the mission of the company, values, and history and assure you can speak to the position you are applying.

2. **Select Relatable Military Experiences:** Prepare for the interview, just as you would prepare for a mission. Craft relatable stories around questions focused on: leadership, ethics, measurable results, global experience, and people management.

3. **Use the "I" Pronoun:** Consciously work on presenting your experience using the pronoun "I."

4. **Attire:** Purchase professional business attire or attire that is appropriate to the industry you wish to enter.

5. **Conversational Dialogue:** Become comfortable in conversational dialogue. Often, communication in the armed services comes off as directive, despite your intent not to be.

6. **Relatable Examples:** Scale your answers when citing military examples. The average person will have difficulty relating an example that starts with I influenced the economy, safety and security, and political stability of country X through the following operations. Answers must be finite and relatable.

7. **Don't be the 'Scary Veteran":** Become accustomed to shaking hands without a "death grip," smile, assume a relaxed posture; in short, be relatable to the person with whom you are speaking.

8. **Compensation, NOT NOW**: NEVER, NEVER discuss compensation during an interview; that is the final piece that can be brough up if an offer is presented.

9. **Thoughtful questions:** Prepare 2-4 thoughtful questions. You will be asked, "what questions do you have?" Examples include:

   a. I see that integrity is a core value of the company. It is also one of my values. Can you discuss how this value is operationalized within the company?

   b. During my time in the Army, I was part of an environment that demonstrates a great deal of care for the members, what does employee care look like in your company?

10. **Critique:** Do not ask the interviewer to critique your interview skills.

11. **Follow up & Gratitude:** At bare minimum, send a thank you email or take a bit of time and mail a handwritten thank you note,

Follow Up and Stay Top of Mind

1. Sending a professional handwritten thank you note to each person you have met with, and in this note based upon the exit reflections of that submission or interview, identify one more compelling better, faster, different, cost-effective key performance indicator and this differentiator about you to share back with them about yourself.

2. Maybe, do the same for anyone within the organization that you did not meet that could have a positive influence on your name moving forward as a "THANK YOU for the opportunity to interview with your organization"

3. Include this outreach follow-up to anyone on your advisory board that assisted you in this interview and anyone inside

the organization that has served as a mentor, advisor, sponsor, or influencer for you.

4. Go online and maybe make a post on LinkedIn about your great interview experience with organization "X" and make their business name a hot link @ in your post

5. Send a Message to anyone you are connected to on LinkedIn about your recent application or interview and ask them for any next step suggestions?

6. Be proactive but not a pest. Follow up with those that you met and interviewed with, if in fact you are excited with the employment aspects, on their LinkedIn account.

7. Have any advocates you have potentially follow up with the organization on your behalf as well, if appropriate.

8. Know when the follow up date and time is and reach out accordingly as your next step connection.

9. Regularly look at your LinkedIn profile administration page, at the top right side is a "Who's viewed your profile" option. Click on that and review the people that have been checking you out. If any look like leads for jobs, a person from an organization that you have applied, or is someone you see from reviewing their profile looks like a great CONNECT, send a CONNECT invite and message.

10. If you have not already been doing so, now is definitely the time to be FOLLOWING, even if you are not CONNECTED, with all the key influencers, contacts, and senior leaders to that firm and every firm you are or have interviewed with. Look at their regular posts and as appropriate COMMENT upon them. Look for opportunities to add valuable comments to their posts.

11. Any valuable market research or intelligence you can do or come across that may be of value to anyone you met, that could serve as another Brand You differentiator and reason

for you to reconnect. Send that information to them as a mere FYI ... This keeps your name top of their mind!

And, for any interview and application submitted that you do not progress forward with, always send a professional follow-up "Thank You for the opportunity to have met and be considered, I greatly appreciate the opportunity that you provided to me to meet, learn, and submit my credentials." Sort of a follow up; always take the high road and never become discouraged.

Other Resources

Do your research on both the industry and business that you are going to be interviewing with for their historical and current trends, fads, norms, expectations, and how you can present yourself as a value-added talent asset to hire amidst everyone else that may also be applying and sitting in the interview cue with you.

The internet is filled with advice on how to best interview. In addition to this chapter, we suggest you look to some of these resources:

- Explore *YouTube* for current and vintage suggestions on how to interview; what questions you should be prepared for; what questions you should ask; what to expect in your first and subsequent interviews; watch and listen to sample examples from experts.

- Explore online *TEDx Talks* for current and vintage suggestions on how to interview; what questions you should be prepared for; what questions you should ask; what to expect in your first and subsequent interviews; watch and listen to sample examples from experts.

- Explore ***SHRM/Society of Human Resource Management*** website, as the professional trade association for HR Professionals. Review their website portal for any insights on current trends for employment, vetting, assessments, ideas that the person on the other side of the video interview or desk may be indoctrinated into – be prepared.

- Follow Best-Selling authors, podcasters, bloggers, and consultants on the ***art and science of interviewing***.

- Follow best-selling authors, podcasters, bloggers, and consultants within the specific industry or business segmentation you are interested in as there may be valuable nuances on the ***art and science of interviewing***.

Call-to-Action:

Now reflect on what you have just read, ideas that have been activated in your head, and weigh them against each element of the **VALOR™ Model** and what you need to be doing right now at this moment in your career/professional life and with what makes sense for you ...

VALOR: The Ultimate Handbook on Military Transition

What are you doing and need to be doing with respect to the strategies and tactics to raise your **VISIBILITY**,

_____

_____

_____

Illustrate ways for you to gain **AWARENESS** into the marketplace,
How can you **LEVERAGE** yourself for your next **OPPORTUNITY**,

_____

_____

_____

While being strategic in the **RELATIONSHIPS** you have now, who are they,

Kevin Preston & Dr. Jeffrey Magee

_____

_____

_____

_____

And who can you identify that you don't know and need to get connected to for potential advocacy in your later career,

_____

_____

_____ !

Chapter 11

You Got The Job!

First, it's time to celebrate. You've worked so hard throughout this process, and you did it! You got the job. Now, it's time to ask yourself, "What's next?"

Is landing your dream job within the industry and company you desire the end or merely a step along a journey? Many of the steps you undertook to be selected for your new and exciting position are ones you will repeat throughout your next career. You must recognize that the career life cycle management of the military is quite different from the corporate world.

A tremendous part of military service is the clearly defined path of promotions. At any point, you are aware of time in grade/service and education to be considered for the next rank. It is transparent and linear. The promotion system outside of the military may not be as clear. What is clear? The tasks you undertook to gain the job are ones you will continue to use to advance your career. Before you start to look ahead for the next opportunity within your new company, you must first learn the organization, culture, and become exceptionally good at the position you were hired for. That is the foundation you will build for advancement.

Now is the time to take everything that you have read in this book and accelerate it forward in this new life. Consider re-looping the entire VALOR process now and focus on deploying it internally to your new employer and organization. Bring maximum value to every individual and business unit possible by applying the VALOR model internally as your next campaign. Consider:

1. Upgrade your mental board of directors with new people who can bring value to you and to whom you can bring value to.

2. Look at your FIST Factor and add new power personalities into each of these five categories.

3. Revisit each letter within the Player capability Index.

4. Continuously upgrade your knowledge, skills, and abilities (KSAs) within your business unit, through any appropriate outside trade associations, and volunteer strategically within your community.

5. Invite opportunities beyond what the organization may have structurally created for performance feedback. If your employer does once a year or twice a year structured performance reviews, look for opportunities to have these one-on-one discussions with your boss much more regularly. If appropriate, seek out senior leaders and/or owners for the same annual one-on-one conversations around your performance, their feedback and insights to where they want to take the business and explore yet-identified opportunities that you may be best positioned to volunteer for or answer.

6. Look for internal and external opportunities to volunteer where you can showcase your greatness and network to meet new people within your industry and company.

7. Leverage the Trajectory Code v-diagram model presented earlier and recognize what are the stepping stones along the

AC Trajectory for success within your new position and team and what are the same answers for a long healthy career within the new organization.

8. Keep your LinkedIn profile updated and relevant. Your future success will be based upon BRAND YOU and how the world can find you and get to know you!

Call-to-Action:

Now reflect on what you have just read, ideas that have been activated in your head, and weigh them against each element of the **VALOR™ Model** and what you need to be doing right now at this moment in your career/professional life and with what makes sense for you ...

VALOR: The Ultimate Handbook on Military Transition

What are you doing and need to be doing with respect to the strategies and tactics to raise your **VISIBILITY,**

_____

_____

_____

Illustrate ways for you to gain **AWARENESS** into the marketplace,
How can you **LEVERAGE** yourself for your next **OPPORTUNITY,**

_____

_____

_____

While being strategic in the **RELATIONSHIPS** you have now, who are they,

_____

_____

_____

And who can you identify that you don't know and need to get connected to for potential advocacy in your later career,

_____

_____

_____ !

Conclusion

You made it! Congratulations on embarking on this journey with us. We are grateful you've entrusted us with your career and your future. Before we leave you, here are some quick tips to use in your new career:

1. Listen far more than you speak; despite your years of military experience, you are brand new. Take time to learn relationships, hierarchies, hot buttons, and how work is accomplished.

2. Drop your Military Rank; your rank is a title of the past. Most people don't understand that title; you will likely address each other by a first name.

3. Don't be the one who consistently states, "When I was in the (insert branch), we did it this way." That gets very old very quickly.

4. Don't be the "scary veteran;" military service is a serious profession as it should be. That often does not translate to your next career. Remember smile, engage in small talk, avoid direct blunt communication, and phrase what could be statements as questions.

5. In the business world, far more than the military, evolution and innovation are the ingredients to success. Always be attentive for opportunities to suggest, participate in, or even lead causes, events, projects, committees, or groups that will allow you to showcase your Brand You talents and elevate those around you.

6. Have fun and enjoy your new adventure.

We'd love for you to keep in touch. Follow both of us on LinkedIn for daily and weekly VALOR inspirational posts, vlogs, blogs and updates -

(94) Kevin Preston, MBA, M Ed | LinkedIn

(94) Dr. Jeffrey Magee ◆ Advisor/Speaker/Author | LinkedIn

You can also explore our many professional development opportunities at both www.JeffreyMagee.com and www.ProfessionalPerformanceMagazine.com.

Remember, in your new career, take any opportunity to meet someone (build those relationships); update your resume with new responsibilities and experiences; and take on any task assigned and do it with joy! Don't leave your military experience behind, care for people on your team, live the values of your branch, and be vulnerable if you don't understand something.

Now it is time to step out and step into your next career. Few individuals get to reinvent themselves; you are living that opportunity. And one last item, pay your knowledge and experiences forward by helping another when they are seeking employment.

~Notes, Doodles & Action Steps~

_____

_____

_____

_____

_____

_____

_____

_____

_____

_____

_____

_____

_____

_____

_____

_____

_____

_____

_____

_____

_____

_____

_____